QUESTIONING TECHNOLOGY

Tool, Toy or Tyrant?

Edited by
John Zerzan & Alice Carnes

New Society Publishers

Philadelphia, PA   Santa Cruz, CA

Gabriola Island, BC

ACKNOWLEDGEMENTS FROM THE EDITORS:
Our thanks to Bob Brubaker for help in formulating the questions that
organize this book and to Howard Besser and David Watson for their help in
finding the excerpts to answer those questions.

Inquires regarding requests to reprint all or part of *Questioning Technology:
Tool, Toy or Tyrant?* should be addressed to:
New Society Publishers
4527 Springfield Avenue
Philadelphia, PA 19143

ISBN 0-86571-204-2 Hardcover
ISBN 0-86571-205-0 Paperback
ISBN Canadian 1-55092-148-7 Hardcover
ISBN Canadian 1-55092-149-5 Paperback

Printed in the United States of America on partially recycled paper by
McNaughton & Gunn, Saline, MI.

Cover design by Tina Birky

To order directly from the publisher, add $1.75 to the price of the first copy,
50¢ each additional and send your check or money order to either:
New Society Publishers   New Society Publishers
PO Box 582   PO Box 189
Santa Cruz, CA 95061   Gabriola Island, BC V0R 1X0

New Society Publishers is a project of the New Society Educational
Foundation, a nonprofit, tax-exempt, public foundation. Opinions expressed
in this book do not necessarily represent positions of the New Society
Educational Foundation.

QUESTIONING TECHNOLOGY, QUESTIONING PATRIARCHY

It is a truism that humans have and will always use tools. Just as obvious, or so it seems to me, is that technology— the use of tools— occurs in a social, political, cultural and economic context, and is never neutral. Tools are always shaped by their use, by the people or institutions which control their production and distribution, and by a culture which validates, circumscribes, or discourages their creation and/or use in different circumstances.

The reverse is of course true as well. Technology and technological decisions structure our minds and, in doing so, our relations with each other and with the natural world. Whether we use tools to control or enhance all our relations is shaped as much by the tools themselves as by any other set of cultural assumptions or social structures.

QUESTIONING TECHNOLOGY helps us identify and trace the threads of technology which are woven into the fabric of our daily lives. It helps us explore the complexity of our personal, political and spiritual interactions with technology; it helps us expose the habits of technological thought which are leading us so quickly and almost unconsciously to environmental disaster; and it helps us examine the mass production and military technologies which are so devastating to so many of our lives.

More crucially, QUESTIONING TECHNOLOGY challenges us to

re-engage our hearts and minds in the search for truly appropriate and accountable technologies. If we are to live as part of an harmonious web of social and ecological relations, we must (re)discover ways to work with other people and species, finding tools which enhance these relations, and renouncing those which threaten the web, whatever their short-term "benefit" to ourselves. We must shift not only our attitudes and actions, but through conscientious decisions concerning our tools, our very habits of thought.

Sadly, there are precious few models to guide those of us who would respond to this challenge. Native, traditional and organic farmers may well have the most to teach in the ongoing work of reconstituting technology in harmony with local communities and the earth. This use of the land demands an attentive awareness of the natural world, patience, a large dose of humility and a stringent accountability to the land, to natural cycles and to the larger human community. To be sure, there's plenty of room for ingenuity, but always within an explicitly cultural, human and natural—not mereley an economic or technological— context.

●   ●   ●

It is ironic that in a book dedicated to questioning both the technology we use and the habits of mind behind today's dominant technological forms, there are many articles which exhibit another, similarly pernicious reality and habit of thought: subsuming all experience under men's experience. Not only does this render a huge majority of people (women especially, but often also children and men not of the dominant culture) invisible, it also robs everyone of the fascinating complexity of experience and perspective which comes with attention to the wonderful diversity of people and cultures.

In this case, many would argue that there is some justice

in the identity of industrial and military technologies with men. These technologies are overwhelmingly invented and controlled by Caucasian men; often their initial victims are women, children and poorer men from other than the dominant culture. Beyond this, many of the habits of thought identified and challenged by feminists are similar to the ones exposed in QUESTIONING TECHNOLOGY.

It is also true, however, that many of the contributors who use the generic male wrote in times when the pattern was less noticed — at least by their colleagues and publishers who were, overwhelmingly, men. Like technology, patriarchy is insidious.

Examples from more recent publications suggest that the same publishing combines which treat their employees as mere parts of their machines also care little about the words — and even (and one is sometimes tempted to say, especially) the ideas — of the authors they publish, as long as the books earn their share of profit for the company.

Facing as we do limited economic and technological choices, we at New Society Publishers have decided not to redo the work of each of the original publishers by contacting them, their editors and each of the authors (some of them deceased) in an attempt to highlight by hand the places where we think changes are necessary.

By adding a bit of hand work to this mass-produced item, we hope to humanize it just a little, and to enhance its challenge to rethink — and remake — our relationships with technology and with you, our community.

TL Hill
for New Society Publishers

CONTENTS

PART ONE
TECHNOLOGY: ITS HISTORY AND OUR FUTURE

PART TWO
COMPUTERS AND THE INFORMED INDIVIDUAL

PART THREE
TECHNOLOGY: THE WEB OF LIFE?

INTRODUCTION

This book presents only one side — the *other* side. Our purpose is to persuade you to think critically about technology.

Forty years ago, no-one thought much about smoking; Ronald Reagan with his Chesterfields was just one instance of the glamour and sophistication associated with cigarettes. Now none but nonentities appear in cigarette ads, and they only on the back covers of magazines, the advertisements having been banned from television. A lot of thought has gone into smoking since a Stetsoned Reagan told its praises, and cigarettes have been officially branded dangerous. Many people have given them up, or are trying hard to do so; almost no-one, apart from tobacco company apologists, promotes cigarette smoking as a good thing.

Computers and automation are now being sold in much the same way that cigarettes were purveyed three or four decades ago. The promise of profit sounds a clarion call to a society reeling from the social and economic dislocations of the past twenty years. Technology is sold as a means of control. Get behind the wheel, run your fingers over the user-friendly keyboard, bark numbers into a headset mike, press a button and watch spreadsheets lay out their graphs. Give in to change, fork over some cash, or better yet, use plastic money — be the emperor of neon green, put yourself in the driver's seat. Be a tamed Charlie Chaplin, at peace with modern times.

Questioning Technology is a purgative, an antidote to the transparent bunkum trafficked by techno wallahs, from the US Army ("technology is taking over the world") to Jazz by Lotus ("You made me love you; I didn't want to do it.") To question technology is to look askance at what our lives are made of, to wonder how our cultural experience has distorted our vision and deformed our human nature. We may be empty of answers and alternatives, but we are full, suddenly, of the need to search them out.

In short, this book is intended to make you mad, wake you up, and get you thinking and talking. Heaving it at the wall is an expected response. Unplugging your computer (or unplugging yourself from one) may be an unexpected outcome, one with happy results for your personal life, health, and future well-being.

You can close the book now . . . and go right on for the next 40 years, smoking your way into the cancer ward. Or you can turn the page . . .

I

TECHNOLOGY
It's History and our Future

---

1. How has technology developed — or encroached? Are the computer, nuclear power, and recombinant DNA comparable to the wheel, the printing press, and gunpowder — or do they represent an entirely new order?

> "The interrelationships of the Greek *tekhne*, a manual skill, and the Greek *tekton*, a carpenter, with the Latin *tegere*, to cover, and *texere*, to weave, are not entirely clear; but that the two Latin words are related to each other, and the Greek to the Latin, can hardly be doubted. Semantically, building (literal and figurative) is either stated or implied in all three groups; phonetically, the relationship is abundantly clear." (Eric Partridge, *Origins: A Short Etymological Dictionary of Modern English*)

Aside from the occasional cataclysm of earthquake or landslide, the earth changes slowly, nearly imperceptibly. Most animal populations make do with what they find and leave their habitats little altered. The few insect, fish, bird, and animal species that build nests, bowers, and other created forms are fascinating to humans, as are the remains of archaic and historic human building.

Technology changes the earth by building. With technology, we select materials abstractly from a variety of settings and

11

meld them into an artificial whole — then sit back to see what will happen. Technology is an impulse, a thought form, before it has anything to do with tools. It grows from the desire to rival the awesome, unfathomable creativity of the earth. This is where domination of nature begins.

In this section, Lewis Mumford discusses democratic and anti-democratic modes of technology, showing the profound impact of what is often overlooked in "political" discourse. Computer pioneer Weizenbaum asks thoughtful questions about the relationship of technical progress to human values, and Michael Shallis also poignantly wonders whether a totally new technological order is upon us.

Mumford is one of many writers who have distinguished between the balanced, bounded technologies of preliterate peoples and the rampant destruction brought on by modern factory industrialism. As you read these selections, make up your own mind: is the difference between archaic and modern technology a difference of degree, or of kind?

Authoritarian and Democratic Technics

LEWIS MUMFORD

"Democracy" is a term now confused and sophisticated by indiscriminate use, and often treated with patronizing contempt. Can we agree, no matter how far we might diverge at a later point, that the spinal principle of democracy is to place what is common to all{men}*above that which any organization, institution, or group may claim for itself? This is not to deny the claims of superior natural endowment, specialized knowledge, technical skill, or institutional organization: all these may, by democratic permission, play a useful role in the human economy. But democracy consists in giving final authority to the whole, rather than the part; and only living human beings, as such, are an authentic expression of the whole, whether acting alone or with the help of others.

Around this central principle clusters a group of related ideas and practices with a long foreground in history, though they are not always present, or present in equal amounts, in all societies. Among these items are communal self-government, free communication as between equals, unimpeded access to the common store of knowledge, protection against arbitrary external controls, and a sense of individual moral responsibility for behavior that affects the whole community. All living organisms are in some degree autonomous, in that they follow a life-pattern of their own; but in {man} this autonomy is an essential condition for {his} further development. We surrender some of our autonomy when ill or crippled: but to surrender it every day on every occasion would be to turn life itself into a chronic illness. The best life possible — and here I am consciously treading on contested ground — is one that calls for an ever greater degree of self-direction, self-expression, and self-realization. In this sense, personality, once the exclusive attribute of kings, belongs on democratic theory to every{man.}Life itself in its fullness and wholeness cannot be delegated.

In framing this provisional definition I trust that I have not, for the sake of agreement, left out anything important. Democracy, in

13

* See "Questioning Technology, Questioning Patriarchy," p.1.

the primal sense I shall use the term, is necessarily most visible in relatively small communities and groups, whose members meet frequently face to face, interact freely, and are known to each other as persons. As soon as large numbers are involved, democratic association must be supplemented by a more abstract, depersonalized form. Historic experience shows that it is much easier to wipe out democracy by an institutional arrangement that gives authority only to those at the apex of the social hierarchy than it is to incorporate democratic practices into a well-organized system under centralized direction, which achieves the highest degree of mechanical efficiency when those who work it have no mind or purpose of their own.

The tension between small-scale association and large-scale organization, between personal autonomy and institutional regulation, between remote control and diffused local intervention, has now created a critical situation. If our eyes had been open, we might long ago have discovered this conflict deeply embedded in technology itself.

I wish it were possible to characterize technics with as much hope of getting assent, with whatever quizzical reserves you may still have, as in this description of democracy. But the very title of this paper is, I confess, a controversial one; and I cannot go far in my analysis without drawing on interpretations that have not yet been adequately published, still less widely discussed or rigorously criticized and evaluated. My thesis, to put it bluntly, is that from late neolithic times in the Near East, right down to our own day, two technologies have recurrently existed side by side: one authoritarian, the other democratic, the first system-centered, immensely powerful, but inherently unstable, the other {man-centered} relatively weak, but resourceful and durable. If I am right, we are now rapidly approaching a point at which, unless we radically alter our present course, our surviving democratic technics will be completely suppressed or supplanted, so that every residual autonomy will be wiped out, or will be permitted only as a playful device of government, like national balloting for already chosen leaders in totalitarian countries.

The data on which this thesis is based are familiar; but their significance has, I believe, been overlooked. What I would call democratic technics is the small scale method of production, resting mainly on human skill and animal energy but always, even when employing machines, remaining under the active direction of the craftsman or the farmer, each group developing its own gifts, through appropriate arts and social ceremonies, as well as making discreet use of the gifts of nature. This technology had limited horizons of achievement, but, just because of its wide diffusion and its modest demands, it had great powers of adaptation and recuperation. This democratic technics has underpinned and firmly supported every historic culture until our own day, and redeemed the constant tendency of authoritarian technics to misapply its powers. Even when paying tribute to the most oppressive authoritarian regimes, there yet remained within the workshop or the farmyard some degree of autonomy, selectivity, creativity. No royal mace, no slave-driver's whip, no bureaucratic directive left its imprint on the textiles of Damascus or the pottery of fifth century Athens.

If this democratic technics goes back to the earliest use of tools, authoritarian technics is a much more recent achievement: it begins around the fourth millennium B.C. in a new configuration of technical invention, scientific observation, and centralized political control that gave rise to the peculiar mode of life we may now identify, without eulogy, as civilization. Under the new institution of kingship, activities that had been scattered, diversified, cut to the human measure, were united on a monumental scale into an entirely new kind of theological-technological mass organization. In the person of an absolute ruler, whose word was law, cosmic powers came down to earth, mobilizing and unifying the efforts of thousands of {men} hitherto all-too-autonomous and too decentralized to act voluntarily in unison for purposes that lay beyond the village horizon.

The new authoritarian technology was not limited by village custom or human sentiment: its herculean feats of mechanical organization rested on ruthless physical coercion, forced labor and slavery, which brought into existence machines that were capable of exerting thousands of horsepower centuries before horses were harnessed or wheels invented. This centralized technics drew on inventions and scientific discoveries of a high order: the written record, mathematics and astronomy, irrigation and canalization: above all, it created complex human machines composed of specialized, standardized, replaceable, interdependent parts — the work army, the military army, the bureaucracy. These work armies and military armies raised the ceiling of human achievement: the first in mass construction, the second in mass destruction, both on a scale hitherto inconceivable. Despite its constant drive to destruction, this totalitarian technics was tolerated, perhaps even welcomed, in home territory, for it created the first economy of controlled abundance: notably, immense food crops that not merely supported a big urban population but released a large trained minority for purely religious, scientific, bureaucratic, or military activity. But the efficiency of the system was impaired by weaknesses that were never overcome until our own day.

To begin with, the democratic economy of the agricultural village resisted incorporation into the new authoritarian system. So even the Roman Empire found it expedient, once resistance was broken and taxes were collected, to consent to a large degree of local autonomy in religion and government. Moreover. as long as agriculture absorbed the labor of some 90 per cent of the population, mass technics were confined largely to the populous urban centers. Since authoritarian technics first took form in an age when metals were scarce and human raw material, captured in war, was easily convertible into machines, its directors never bothered to invent inorganic mechanical substitutes. But there were even greater weaknesses: the system had no inner coherence: a break in communication, a missing link in the chain of command, and the great human machines fell apart. Finally, the myths upon which the whole system was based — particularly the essential myth of kingship — were irrational, with their paranoid suspicions and animosities and their paranoid claims to unconditional obedience and absolute power. For all its redoubtable constructive achievements, authoritarian technics expressed a deep hostility to life.

By now you doubtless see the point of this brief historic excursus. That authoritarian technics has come back today in an immensely magnified and adroitly perfected form. Up to now, following the optimistic premises of nineteenth century thinkers like Auguste Comte and Herbert Spencer, we have regarded the spread of experimental science and mechanical invention as the soundest guarantee of a peaceful, productive, above all democratic, industrial society. Many have even comfortably supposed that the revolt against arbitrary political power in the seventeenth century was causally connected with the industrial revolution that accompanied it. But what we have interpreted as the new freedom now turns out to be a much more sophisticated version of the old slavery: for the rise of political democracy during the last few centuries has been increasingly nullified by the successful resurrection of a centralized authoritarian technics — a technics that had in fact for long lapsed in many parts of the world.

Let us fool ourselves no longer. At the very moment Western nations threw off the ancient regime of absolute government, operating under a once-divine king, they were restoring this same system in a far more effective form in their technology, reintroducing coercions of a military character no less strict in the organization of a factory than in that of the new drilled, uniformed, and regimented army. During the transitional stages of the last two centuries, the ultimate tendency of this system might be in doubt, for in many areas there were strong democratic reactions; but with the knitting together of a scientific ideology, itself liberated from theological restrictions or humanistic purposes, authoritarian technics found an instrument at hand that has now given it absolute command of physical energies of cosmic dimensions. The inventors of nuclear bombs, space rockets, and computers are the pyramid builders of our own age: psychologically inflated by a similar myth of unqualified power, boasting through their science of their increasing omnipotence, if not omniscience, moved by obsessions and compulsions no less irrational than those of earlier absolute systems: particularly the notion that the system itself must be expanded, at whatever eventual cost to life.

Through mechanization, automation, cybernetic direction, this authoritarian technics has at last successfully overcome its most serious weakness: its original dependence upon resistant, sometimes actively disobedient servo-mechanisms, still human enough to harbor purposes that do not always coincide with those of the system.

Like the earliest form of authoritarian technics, this new technology is marvellously dynamic and productive: its power in every form tends to increase without limits, in quantities that defy assimilation and defeat control, whether we are thinking of the output of scientific knowledge or of industrial assembly lines. To maximise energy, speed, or automation, without reference to the complex conditions that sustain organic life, have become ends in themselves. As with the earliest forms of authoritarian technics, the weight of effort, if one is to judge by national budgets, is toward absolute instruments of destruction, designed for absolutely irrational purposes whose chief by-product would be the mutilation or extermination of the human race. Even Ashurbanipal and

Genghis Khan performed their gory operations under normal human limits.

The center of authority in this new system is no longer a visible personality, an all-powerful king: even in totalitarian dictatorships the center now lies in the system itself, invisible but omnipresent: all its human components, even the technical and managerial elite, even the sacred priesthood of science, who alone have access to the secret knowledge by means of which total control is now swiftly being effected, are themselves trapped by the very perfection of the organization they have invented. Like the pharoahs of the Pyramid Age, these servants of the system identify its goods with their own kind of well-being: as with the divine king, their praise of the system is an act of self-worship; and again like the king, they are in the grip of an irrational compulsion to extend their means of control and expand the scope of their authority. In this new systems-centered collective, this Pentagon of power, there is no visible presence who issues commands: unlike Job's God, the new deities cannot be confronted, still less defied. Under the pretext of saving labor, the ultimate end of this technics is to displace life, or rather, to transfer the attributes of life to the machine and the mechanical collective, allowing only so much of the organism to remain as may be controlled and manipulated.

Do not misunderstand this analysis. The danger to democracy does not spring from any specific scientific discoveries or electronic inventions. The human compulsions that dominate the authoritarian technics of our own day date back to a period before even the wheel had been invented. The danger springs from the fact that, since Francis Bacon and Galileo defined the new methods and objectives of technics, our great physical transformations have been effected by a system that deliberately eliminates the whole human personality, ignores the historic process, overplays the role of the abstract intelligence, and makes control over physical nature, ultimately control over {man}{himself,} the chief purpose of existence. This system has made its way so insidiously into Western society, that my analysis of its derivation and its intentions may well seem more questionable — indeed more shocking — than the facts themselves.

Why has our age surrendered so easily to the controllers, the manipulators, the conditioners of an authoritarian technics? The answer to this question is both paradoxical and ironic. Present day technics differs from that of the overtly brutal, half-baked authoritarian systems of the past in one highly favorable particular: it has accepted the basic principle of democracy, that every member of society should have a share in its goods. By progressively fulfilling this part of the democratic promise, our system has achieved a hold over the whole community that threatens to wipe out every other vestige of democracy.

The bargain we are being asked to ratify takes the form of a magnificent bribe. Under the democratic-authoritarian social contract, each member of the community may claim every material advantage, every intellectual and emotional stimulus he may desire, in quantities hardly available hitherto even for a restricted minority: food, housing, swift transportation, instantaneous communication, medical care, entertainment, education. But on one condition: that one must not merely ask for nothing that the system does not provide, but likewise agree to take everything offered, duly processed and fabricated, homogenized and equalized, in the precise quantities that the system, rather than the person, requires. Once one opts for the system no further choice remains. In a word, if one surrenders one's life at source, authoritarian technics will give back as much of it as can be mechanically graded, quantitatively multiplied, collectively manipulated and magnified.

"Is this not a fair bargain?" those who speak for the system will ask. "Are not the goods authoritarian technics promises real goods? Is this not the horn of plenty that mankind has long dreamed of, and that every ruling class has tried to secure, at whatever cost of brutality and injustice, for itself?" I would not belittle, still less deny, the many admirable products this technology has brought forth, products that a self-regulating economy would make good use of. I would only suggest that it is time to reckon up the human disadvantages and costs, to say nothing of the dangers, of our unqualified acceptance of the system itself. Even the immediate price is heavy; for the system is so far from being under effective human direction that it may poison us wholesale to provide us with food or exterminate us to provide national security, before we can enjoy its promised goods. Is it really humanly profitable to give up the possibility of living a few years at Walden Pond, so to say, for the privilege of spending a lifetime in *Walden Two*? Once our authoritarian technics consolidates its powers, with the aid of its new forms of mass control, its panoply of tranquillizers and sedatives and aphrodisiacs, could democracy in any form survive? That question is absurd: life itself will not survive, except what is funneled through the mechanical collective. The spread of a sterilized scientific intelligence over the planet would not, as Teilhard de Chardin so innocently imagined, be the happy consummation of divine purpose: it would rather ensure the final arrest of any further human development.

Again: do not mistake my meaning. This is not a prediction of what *will* happen, but a warning against what *may* happen.

What means must be taken to escape this fate? In characterizing the authoritarian technics that has begun to dominate us, I have not forgotten the great lesson of history: Prepare for the unexpected! Nor do I overlook the immense reserves of vitality and creativity that a more humane democratic tradition still offers us. What I wish to do is to persuade those who are concerned with maintaining democratic institutions to see that their constructive efforts must include technology itself. There, too, we must return to the human center. We must challenge this authoritarian system that has given to an underdimensioned ideology and technology the authority that belongs to the human personality. I repeat: life cannot be delegated.

Curiously, the first words in support of this thesis came forth, with exquisite symbolic aptness, from a willing agent — but very nearly a classic victim! — of the new authoritarian technics. They came from the astronaut, John Glenn, whose life was endangered by the malfunctioning of his automatic controls, operated from a remote center. After he barely saved his life by personal intervention, he emerged from his space capsule with these ringing words: "Now let man take over!"

That command is easier to utter than obey. But if we are not to be driven to even more drastic measures than Samuel Butler suggested in *Erewhon*, we had better map out a more positive course: namely, the reconstitution of both our science and our technics in such a fashion as to insert the rejected parts of the human personality at every stage in the process. This means gladly sacrificing mere quantity in order to restore qualitative choice, shifting the seat of authority from the mechanical collective to the human personality and the autonomous group, favoring variety and ecological complexity, instead of stressing undue uniformity and standardization, above all, reducing the insensate drive to extend the system itself, instead of containing it within definite human limits and thus releasing man himself for other purposes. We must ask, not what is good for science or technology, still less what is good for General Motors or Union Carbide or IBM or the Pentagon, but what is good for {man:} not machine-conditioned, system-regulated, {mass-man,} but {man} in person, moving freely over every area of life.

There are large areas of technology that can be redeemed by the democratic process, once we have overcome the infantile compulsions and automatisms that now threaten to cancel out our real gains. The very leisure that the machine now gives in advanced countries can be profitably used, not for further commitment to still other kinds of machine, furnishing automatic recreation, but by doing significant forms of work, unprofitable or technically impossible under mass production: work dependent upon special skill, knowledge, aesthetic sense. The do-it-yourself movement prematurely got bogged down in an attempt to sell still more machines; but its slogan pointed in the right direction, provided we still have a self to do it with. The glut of motor cars that is now destroying our cities can be coped with only if we redesign our cities to make fuller use of a more efficient human agent: the walker. Even in childbirth, the emphasis is already happily shifting from an officious, often lethal, authoritarian procedure, centered in hospital routine, to a more human mode, which restores initiative to the mother and to the body's natural rhythms.

The replenishment of democratic technics is plainly too big a subject to be handled in a final sentence or two: but I trust I have made it clear that the genuine advantages our scientifically based technics has brought can be preserved only if we cut the whole system back to a point at which it will permit human alternatives, human interventions, and human destinations for entirely different purposes from those of the system itself. At the present juncture, if democracy did not exist, we would have to invent it, in order to save and recultivate the spirit of {man.}

Computer Power
and Human Reason

JOSEPH WEIZENBAUM

This book is only nominally about computers. In an important
sense, the computer is used here merely as a vehicle for moving
certain ideas that are much more important than computers. The
reader who looks at a few of this book's pages and turns away in
fright because he spots an equation or a bit of computer jargon here
and there should reconsider. He may think that he does not know
anything about computers, indeed, that computers are too
complicated for ordinary people to understand. But a major point of
this book is precisely that we, all of us, have made the world too
much into a computer, and that this remaking of the world in the
image of the computer started long before there were any electronic
computers. Now that we have computers, it becomes somewhat
easier to see this imaginative transformation we have worked on the
world. Now we can use the computer itself — that is the idea of the
computer — as a metaphor to help us understand what we have
done and are doing. . . .

"Concepts have been reduced to summaries of the characteristics that
several specimens have in common. By denoting similarity, concepts
eliminate the bother of enumerating qualities and thus serve better to
organize the material of knowledge. They are thought of as mere
abbreviations of the items to which they refer. Any use transcending
auxiliary, technical summarization of factual data has been eliminated as a
last trace of superstition. Concepts have become 'streamlined', rationalized,
labor-saving devices . . . thinking itself [has] been reduced to the level of
industrial processes . . . in short, made part and parcel of production."[1]

No one who does not know the technical basis of the systems we
have been discussing can possibly appreciate what a chillingly
accurate account of them this passage is. It was written by the

22

*See "Questioning Technology, Questioning Patriarchy," p. 1.

philosopher-sociologist Max Horkeimer in 1947, years before the forces that were even then eclipsing reason, to use Horkeimer's own expression, came to be embodied literally in machines.

This passage, especially in view of when and by whom it was written, informs us once again that the computer, as presently used by the technological elite, is not a cause of anything. It is rather an instrument pressed into the service of rationalizing, supporting, and sustaining the most conservative, indeed, reactionary, ideological components of the current *Zeitgeist*.

As we see so clearly in the various systems under scrutiny, meaning has become entirely transformed into function. Language, hence reason too, has been transformed into nothing more than an instrument for affecting the things and events in the world. Nothing these systems do has any intrinsic significance. There are only goals dictated by tides that cannot be turned back. There are only means-ends analyses for detecting discrepancies between the way things are, the "observed condition", and the way the fate that has befallen us tells us we wish them to be. In the process of adapting ourselves to these systems, we, even the admirals among us, have {castrated} not only ourselves (that is, resigned ourselves to impotence), but our very language as well. For now that language has become merely another tool, all concepts, ideas, images that artists and writers cannot paraphrase into computer-comprehensible language have lost their function and their potency. Forrester tells us this most clearly — but the others can be seen nodding their agreement: "Any concept and relationship that can be clearly stated in ordinary language can be translated into computer model language." The burden of proof that something has been "stated clearly" is on the poet. No wonder, given this view of language, that the distinction between the living and the lifeless, between {man} and machine, has become something less than real, at most a matter of nuance!

Corrupt language is very deeply imbedded in the rhetoric of the technological elite. We have already noted the transformation of the meaning of the word "understand" by Minsky into a purely instrumental term. And it is this interpretation of it that, of course, pervades all the systems we have been discussing. Newell and Simon's use of the word "problem" is another example and one just as significant.

During the times of trouble on American university campuses, one could often hear well-meaning speakers say that the unrest, at least on their campuses, was mainly caused by inadequate communication among the university's various constituencies, e.g., faculty, administration, students, staff. The "problem" was therefore seen as fundamentally a communication, hence a technical, problem. It was therefore solvable by technical means, such as the establishment of various "hotlines" to, say, the president's or the provost's office. Perhaps there were communciation difficulties; there usually are on most campuses. But this view of the "problem" — a view entirely consistent with Newell and Simon's view of "human problem solving" and with instrumental reasoning — actively hides, buries, the existence of real conflicts. It may be, for example, that students have genuine ethical, moral, and political interests that conflict with interests the university administration perceives itself to have, and that each constituency understands the other's interests very well. Then there is a genuine problem, not a communication difficulty, certainly not one that can be repaired by the technical expedient of hotlines. but instrumental reason converts each dilemma, however genuine, into a mere paradox that can then be unraveled by the application of logic, by calculation. All conflicting interests are replaced by the interests of technique alone.

This, like Philip Morrison's story, is a parable too. Its wider significance is that the corruption of the word "problem" has brought in its train the mystique of "problem solving", with catastrophic effects on the whole world. When every problem on the international scene is seen by the "best and brightest" problem solvers as being a mere technical problem, wars like the Viet Nam war become truly inevitable. The recognition of genuinely conflicting but legitimate interests of coexisting societies — and such recognition is surely a precondition to conflict resolution or accommodation — is rendered impossible from the outset. Instead, the simplest criteria are used to detect differences, to search for means to reduce these differences, and finally to apply operators to "present objects" in order to transform them into "desired objects". It is, in fact, entirely reasonable, if "reason" means instrumental reason, to apply American military force, B-52's, napalm, and all the rest, to "communist-dominated" Viet Nam (clearly an "undesirable object"), as the "operator" to transform it into a "desirable object", namely, a country serving American interests.

The mechanization of reason and of language has consequences far beyond any envisioned by the problem solvers we have cited.

Horkeimer, long before computers became a fetish and gave concrete form to the eclipse of reason, gave us the needed perspective:

"Justice, equality, happiness, tolerance, all the concepts that . . . were in preceding centuries supposed to be inherent in or sanctioned by reason, have lost their intellectual roots. They are still aims and ends, but there is no rational agency authorized to appraise and link them to an objective reality. Endorsed by venerable historical documents, they may still enjoy a certain prestige, and some are contained in the supreme law of the greatest countries. Nevertheless, they lack any confirmation by reason in its modern sense. Who can say that any one of these ideals is more closely related to truth than its opposite? According to the philosophy of the average modern intellectual, there is only one authority, namely, science, conceived as the classification of facts and the calculation of probabilities. The statement that justice and freedom are better in themselves than injustice and oppression is scientifically unverifiable and useless. It has come to sound as meaningless in itself as would the statement that red is more beautiful than blue, or that an egg is better than milk."[2] . . .

The theories — or, perhaps better said, the root metaphors — that have hypnotized the artificial intelligentsia, and large segments of the general public as well, have long ago determined that life is what is computable and only that. As Professor John McCarthy, head of Stanford University's Artificial Intelligence Laboratory said, "The only reason we have not yet succeeded in formalizing every aspect of the real world is that we have been lacking a sufficiently powerful logical calculus. I am currently working on that problem."

Sometimes when my children were still little, my wife and I would stand over them as they lay sleeping in their beds. We spoke to each other in silence, rehearsing a scene as old as mankind itself. It is as Ionesco told his journal: "Not everything is unsayable in words, only the living truth."

---

1 M. Horkeimer, *Eclipse of Reason*(New York: Seabury, 1974), p. 21. This important book was first published by the Oxford University Press, New York, in 1947.
2 *Ibid.*, pp. 23-24.

The Silicon Idol

MICHAEL SHALLIS

The interaction of {man} with {his} technologies has transformed the world and has transformed {man}. The extension of {man's} natural senses and abilities, through the development of tools, techniques and the media of communication, has altered nature and {man's} attitude to it as well as reflected it. Incorporated into a technology is a segment of {man's} cosmology, {his} view of the universe, {his} own skills, reasoning and imagination. Built into each and every tool {man} makes, is an extension of {man} {himself} and yet the tool, the extension of {man} into {his} technology, reflects {him} imperfectly, distorts the image and operates on the world and on humanity in ways so different from those intended as to modify and modulate the world unexpectedly.

Combined with the power of technology to transform is its spell-binding power to fascinate. The psalmist's line, "They that make them are like unto them", points out this power and Marshall McLuhan links it to the Narcissus myth. Narcissus as a word comes from the Greek *narcosis*, which also gives us the term narcotic, which means numbness. McLuhan, in *Understanding Media*, wrote:

"The youth Narcissus mistook his own reflection in the water for another person. This extension of himself by mirror numbed his perceptions until he became the servo-mechanism of his own extended or repeated image. The nymph Echo tried to win his love with fragments of his own speech, but in vain. He was numb. He had adapted to his extension of himself and had become a closed system."(p.51)

This passage undoubtedly encapsulates much of the essence of technology. Narcissus' mirror symbolizes all technologies, reflecting man or some aspect of his capacities directly in an external form. We are used to thinking of technology in terms of heavy machinery, things with cogs and wheels, but silicon chip technology is not like that and the above myth illustrates the fact that even the surface of a pool, a natural phenomenon, can become a technology, an extension of {man}. Equally a technique, a process for doing something (maybe looking at one's own reflection), can become an external

26

*See "Questioning Technology, Questioning Patriarchy", p.1.

embodiment of human faculties and equivalent to technology, especially when seen as a thing in itself. This point, too, emerges from McLuhan's paragraph; the trouble with technology is not itself but our attitude to it, for we do not often recognise it for what it is, mistaking it for something that is an end in itself rather than a means to an end. Narcissus misunderstood the meaning of the image he saw in the water; it was a mirror, a means of seeing himself, and the reflection was not a "real" object at all.

The numbness induced in Narcissus, which gives us his name, is also a general reaction induced by technology and by techniques. Travel by motor car numbs one's senses to the reality of the countryside, leaving the traveller exhausted and cut off, in other words numbed. Walking, on the other hand, unless pursued fanatically as a technique to be regarded as an end in itself, may leave the traveller physically tired but the senses, the sense of integration with the world, should be extended not numbed. Travelling on foot can be wholesome in a way that travelling by car (or train or plane) can never be; the technology drugs its user. Narcotic, apart from meaning numbness, also has the connotation of addiction and that, too, is true of technology. The car user uses the car instead of walking, even for journeys of only a few hundred yards. The technical means of 'easy' travel induces its use whenever possible not just when appropriate. The television viewer finds it difficult to switch off after one programme, but tends to see what is on next, just as the smoker has to have one more cigarette. People turn on electric lights because they are there even when natural light is in abundance. Technologies are not easy to switch off or leave aside, so the numbness they induce becomes reinforced.

Narcissus shows us the mythological picture of the hacker, fascinated with the system to the point of obsessiveness, to the exclusion of all else. The numbness induced by the technology reinforces this obsession for it blots out all other influences. The myth may seem to be an extreme case, as does that of the hacker, until we try to think of all the technologies we use in daily life and contemplate alternatives to each of them or the idea of not using any of them. Our own addiction to our extensions in technology is more than a pale shadow of that shown by Narcissus, and the reason for it is also given by McLuhan. He describes the "addiction" in terms of the user of technology becoming a servo-mechanism of the technology, adapting to the technique and becoming thereby closed to the world. {Man} and technology are symbiotic, in that sense, because each feeds off the other; {man} invents the technology, develops an extension of himself and then adapts himself to the device of his own making. The servo-mechanism of such a cycle then prevents the user from returning to old ways because the mechanism of the new technique has excluded the thing it replaced. The feedback loop of the technology and its use can only be broken by willed effort. Such a procedure also manifests itself in the notion of the technological imperative: the idea that technology has a life of its own, what can be done must be done. It is the argument of a Narcissus treating his own reflection as an end in itself.

John Biram coined a term for man's obsession with technology and for the general misunderstanding that technologies and techniques are not ends but means to ends; the term is *teknosis*. In his book of the same name he argues that the teknotic attitude can be found in everyone and is related to so much of modern life that our whole culture suffers chronically from this "disease of technical thinking". Teknosis is an attitude of mind, part hacker syndrome, part Narcissus complex, whereby man worships idols of silver and gold and becomes like the objects of worship. Religious rituals, after all, are designed to help one reach upwards to God, so it is not surprising that the profane rituals of technique help one reach outwards towards one's own mirror image; a direction not likely to be helpful to one spiritually, as the goal is misplaced, or emotionally, as one's affections are directed to an image rather than to something "real". Teknosis implies unreality, which seems a strange notion when our technology, our hardware, seems so *real*, so predominant in all aspects of our physical environment. Teknosis is akin to Heidegger's notion of "technicity", which W. J. Richardson has described thus: "Rather it is the fundamental attitude in man by which all beings, even himself, become raw material for his pro-posing, contra-posing, (self)-imposing compartment with beings. Technology is simply the instrumentation of this attitude (p.44)." The idea that technology is the "hardware" by which man extends his attitudes, beliefs and thoughts is present in all the threads I have drawn together here from McLuhan, Biram and Heidegger. Technology reflects man and man is himself re-reflected by his technology.

This web of technical description and of technical implementation also redefines the world in its own terms. Marshall McLuhan has encapsulated this ability, indeed essential function, of all technologies in the phrase "the medium is the message". The transforming power of techniques is the technique not its content. Printing changed the medieval world dramatically in thought,

practice and outlook, even though most people could not read, and
not many even saw a book. Yet printing transformed the lives of
most people in that society, because the medium of print, that new
technology, changed attitudes, altered perspectives, redefined the
world. It was the fact of print, not what was actually printed, that
transformed society, and McLuhan is right when he comments that
we too often ignore the medium and only see the content. The
technology that, arguably, best illustrates his point is the contentless
technology of electric light. The content is pure illumination and yet
this medium, this technology, transforms man's individual life and
revolutionizes society. All electric technology is pure information
technology where the content is irrelevant. It is most starkly shown
by the electric light but is demonstrated in computing and
communications technology generally.

Electric light illuminates our physical surroundings. The
information it provides comes from our relationship with that
environment. If we are using the light to read a book then the book
becomes the content of the medium of electric light. What the book
is has no relevance to our altered perspective that the light has
induced. The irrelevance of the content is shown more starkly when
the light is turned on to brighten an already adequately lit corridor.
Such usage of a technology also reflects our numbed reliance on it.
How often do we stop to ask ourselves: "Do I need to supplement the
light in this room to see what I am doing now?"

One of the difficulties of discussing the irrelevance of content in
assessing electric and communications technologies is that the
information being communicated gets in the way. The effect of
television as a technology is always masked by the programmes it
has to offer. However, the programmes are irrelevant, the
technology interacts with the viewer in many ways, irrespective of
the content. The technology intervenes between person and
commonly shared experiences. Discussing last night's TV
programme with friends is more fulfilling than watching the
programme in the first place. The important question is not what
information you received but to what degree did the technology
intervene between you and the wholesome experience. In watching
television with friends one finds that although each person is having
identical images and sound presented to them, the experience
violates each individual and does not bring them together into a
common pool of shared human interaction. Electric media,
operating on the central nervous system, engage people at deep

levels of attention, narrowing down the focus of their senses, enmeshing them in contentious webs.

Although the printing press is often regarded as the turning point of medieval society, Lewis Mumford, in *Technics and Civilization*, suggests that the earlier invention of the clock was more crucial as an agent of change. The mechanical clock introduced a linear, progressive, sequential awareness of time, in place of the organic, cyclic perception of time man had before. The clock transformed society and subjected people to the rule of time, to work by the clock, to do things when the church clock struck the hour or quarters, rather than when ready to be done. As I have said in my own book, *On Time*, the mechanization of time, its subjugation to technology, paved the way for the mechanization of speech, through the printing press, and the mechanization of space through modern transport. However, the mechanical clock, as Weizenbaum so rightly pointed out, also introduced another aspect to technology, namely autonomy.

Most technologies extend human faculties. The wheel extends the foot, the telescope extends the eye, our clothes extend our skin. Such technologies, extensions of the human body, can be classed as *prosthetic*, artificial "limbs and organs". The clock, however, does not extend a human faculty, despite the fact that our bodies react to several internal biological "clocks". The clock is an *autonomous* machine, automatic in that it works by itself unaided, once wound up, and independent of man for what it does. The clock embodies a model of planetary motion to provide elements of time, it does not reflect man physically at all, neither does it measure a physical quality, for time is immaterial, but it does reflect man's ideas. The transforming power of the clock lies in its autonomy, its independence of man. It is a mirror of a rational idea.

Lewis Mumford wrote that the clock "dissociated time from human events and helped create the belief in an independent world of mathematically measurable sequences", to which Weizenbaum adds: "the clock created literally a new reality ... that was and remains an impoverished version of the old one". The disassociation, the abstraction of time from human experience, is the consequence of the autonomy of the clock. The medium is bereft of human qualities, independent from nature, and so the message, too, is abstract, objective. It is this objectivity also that impoverishes man's life by placing a framework around society that removes people from their previous more intimate relationship with nature. The grid of time set up mechanically is a harness that reduces the richness of life. That is how all technologies work in one way or another and the inevitable impoverishment they cause must be weighed against the material advantages that their acceptance implies. That impoverishment, that removal from nature, arises with all technology but seems most powerful and embracing with the autonomous clock. The extensions of {man} being of {man} have more subjectivity than does the abstract machine.

If the clock was the first autonomous machine then the second one is the computer. The computer is not prosthetic, it is not an extension of the brain, and, despite Marshall McLuhan's analysis, I do not think that the computer and its extension through electronic communications technology is essentially an extension of the central nervous system. This latter analogy seems plausible but it does imply the brain/computer equation which I do not think is appropriate, although the electric technologies do act on the central nervous system. Rather than extend {man's} own characteristics, the computer, like the clock, encapsulates an abstract notion, it embodies one of {man's} ideas. The computer models the notion of pure rationality, {man's} ideal view of his own intelligence. The impoverishment comes from the limitation of that view of {man} for intelligence is far more than pure rationality. The power of the computer, like the power of the clock, comes from its independence of {man} leaving {man} with {his} Narcissus-like infatuation with {his} own created images, to redefine {himself} in the image of the computer.

Technologies embody {man's} view of the world and {his} relationship with it and hence the machines {man} builds are symbols of {man's} re-created world. As symbols are a means of communication so all machines are communications media and their form is their pervasive message. The anonymity of a computer, its facelessness, symbolizes {man's} own loss of face and faith, {his} own anonymity in a world of machines, a world become teknotic. The network of mechanical time laid down by the clock has been hardened and overlaid by a network of information channels. The clock changed {man} from being a creature in nature to being a master of nature. Electronic microtechnology is changing {man} from being {master} of nature to becoming slave to the autonomous machine. The invisible web of information, the interfaced black boxes, act as agents for change, as do all technologies, but, whereas prosthetic tools embody the designs of their own replacement, the autonomous computer embodies {man's} own replacement by the machine that "thinks".

2. *Was there a point in history when technology came to dominate the individual? How could this have happened?*

In Arcadia, the mythic Golden age, we were naked, beautiful, and unashamed. Nature was neither landscape nor background; we were of the world, not set against it. Human noises blended with bird song, the hum of insects, and the cries of animals.

Tasting the forbidden fruit, opening Pandora's box, stealing fire from the gods — these are the sorts of chosen, willful acts that, in the stories, bring the Golden Age to a catastrophic close.

What truths lie behind these legends? Did everything begin to shift when humans first spoke words, planted seeds, or put each other forcibly to work?

We can only speculate about the origins of technology; the imagined scenario is a mirror of present awareness. This startling reflection displays the accepted world as it really is — arbitrary and imposed. There is nothing inevitable about the technological landscape. It has been chosen, and human participation sustains its reality moment by moment, building an ever higher rampart to keep out the world we once were part of.

Here, anthropologist Stanley Diamond contemplates what we may have lost as technical evolution destroyed traditional society. Historian Carolyn Merchant sees the seventeenth century as the definitive period that established modern — and anti-human — science and technology. Jacques Ellul has offered perhaps the most systematic philosophical indictment of technology's trajectory; here is a representative, essential piece of his argument, excerpted from *The Technological Society*.

In Search of the Primitive

STANLEY DIAMOND

In machine-based societies, the machine has incorporated the demands of the civil power or of the market, and the whole life of society, of all classes and grades, must adjust to its rhythms. Time becomes lineal, secularized, "precious"; it is reduced to an extension in space that must be filled up, and sacred time disappears. The secretary must adjust to the speed of {her}* electric typewriter; the stenographer to the stenotype machine; the factory worker to the line or lathe; the executive to the schedule of the train or plane and the practically instantaneous transmission of the telephone; the chauffeur to the superhighways; the reader to the endless stream of printed matter from high-speed presses; even the {schoolboy} to the precise periodization of {his} day and to the watch on {his} wrist; the person "at leisure" to a mechanized domestic environment and the flow of efficiently scheduled entertainment. The machines seem to run us, crystallizing in their mechanical or electronic pulses the means of our desires. The collapse of time to an extension in space, calibrated by machines, has bowdlerized our natural and human rhythms and helped dissociate us from ourselves. Even now, we hardly love the earth or see with eyes or listen any longer with our ears, and we scarcely feel our hearts beat before they break in protest. Even now, so faithful and exact are the machines as servants that they seem an alien force, persuading us at every turn to fulfill our intentions which we have built into them and which they represent — in much the same way that the perfect body servant routinizes and, finally, trivializes {his} master.

Of such things, actual or possible, primitive societies have no conception. Such things are literally beyond their wildest dreams, beyond their idea of alienation from village or family or the earth itself, beyond their conception of death, which does not estrange them from society or nature but completes the arc of life. There is only one rough analogy. The fear of excommunication from the kinship unit, from the personal nexus that joins {man} society and nature in an endless round of growth (in short, the sense of being isolated and depersonalized and, therefore, at the mercy of demonic

* See "Questioning Technology, Questioning Patriarchy," p. 1.

forces — a fear widespread among primitive peoples) may be taken as an indication of how they would react to the technically alienating processes of civilization if they were to understand them. That is, by comprehending the attitude of primitive people about excommunication from the web of social and natural kinship we can, by analogy, understand their repugnance and fear of civilization.

Primitive society may be regarded as a system in equilibrium, spinning kaleidoscopically on its axis but at a relatively fixed point. Civilization may be regarded as a system in internal disequilibrium; technology or ideology or social organization are always out of joint with each other — that is what propels the system along a given track. Our sense of movement, of *incompleteness*, contributes to the idea of progress. Hence, the idea of progress is generic to civilization. And our idea of primitive society as existing in a state of dynamic equilibrium and as expressive of human and natural rhythms is a logical projection of civilized societies and is in opposition to civilization's actual state. But it also coincides with the real historical condition of primitive societies. The longing for a primitive mode of existence is no mere fantasy or sentimental whim; it is consonant with fundamental human needs, the fulfillment of which (although in different form) is a precondition for our survival. Even the skeptical and civilized Samuel Johnson, who derided Boswell for his intellectual affair with Rousseau, had written:

when {man} began to desire private property then entered violence, and fraud, and theft, and rapine. Soon after, pride and envy broke out in the world and brought with them a new standard of wealth, for {men,} who till then, thought themselves rich, when they wanted nothing, now rated their demands, not by the calls of nature, but by the plenty of others; and began to consider themselves poor, when they beheld their own possessions exceeded by those of their neighbors.

This may be inadequate ethnology, but it was the *cri de coeur* of a civilized {man} for a surcease from mere consumption and acquisitiveness, and so interpreted, it assumes something about primitive societies that is true, namely, predatory property, production for profit does not exist among them.

The search for the primitive is, then, as old as civilization. It is the search for the utopia of the past, projected into the future, with civilization being the middle term. It is birth, death, and transcendent rebirth, the passion called Christian, the trial of Job, the oedipal transition, the triadic metaphor of human growth, felt also in the vaster pulse of history. And this search for the primitive is inseparable from the vision of civilization. No prophet or philosopher of any consequence has spelled out the imperatives of his version of a superior civilization without assuming certain constants in human nature and elements of a primitive condition, without, in short, engaging in the anthropological enterprise. A utopia detached from these twin pillars — a sense of human nature and a sense of the precivilized past — becomes a nightmare. For humanity must then be conceived to be infinitely adaptable and thus incapable of historic understanding or self-amendment. Even Plato's utopia presumes, at least, a good if no longer viable prior state, erroneously conceived as primitive by the refined Greek when it was merely rustic; and the *Republic* was, after all, founded on a theory of human nature that was certainly wrong. Nevertheless, it was a saving grace, for Plato believed that his perfectly civilized society would realize human possibilities not merely manipulate them.

Even the most brilliant and fearful utopian projections have been compelled to solve the problem of the human response, usually with some direct or allegorical reference to a prior or primitive level of functioning. In Zamiatin's *We*, a satirical work of great beauty, the collective society of the future is based on, and has become a maleficent version of, Plato's *Republic*. The people have been reduced to abstract ciphers, their emotions have been controlled and centralized (as in the Republic, mathematics is the most sublime language; but it is not a means of human communication, only an abstract dialogue with God); and history has ceased to exist. Zamiatin documents the growth of the internal rebel who is gradually educated in the experience of what the regime defines as love. When the revolt against this state of happiness occurs, the civil power uses two ultimate weapons: one is a method of instantaneously disintegrating the enemy. Since the enemy is legion, the other method is the "salvation" of the person, as an eternal civil servant, through a quick, efficient operation on the brain that results in a permanent dissociation of intellect and emotion without impairing technical intelligence. Zamiatin's description of the rebel rendered affectless, lucidly describing the changes on his beloved coconspirator's face and feeling nothing as she dies, anticipates Camus and transmits in its terrifying, poignant flatness a psychological truth about our time that has become a dreadful cliché. Zamiatin informs us that such a materialist, secularized and impersonal utopia can function only by altering human nature itself.

And, outside the glass wall of his utopian city which had arisen out of the ruin of the "final" war between the country and the city is a green wilderness in which primitive rebels live off the land, alive to their humanity, and seek to free the ultimately urbanized brother within.

The Death of Nature

CAROLYN MERCHANT

The fundamental social and intellectual problem for the seventeenth century was the problem of order. The perception of disorder, so important to the Baconian doctrine of dominion over nature, was also crucial to the rise of mechanism as a rational antidote to the disintegration of the organic cosmos. The new mechanical philosophy of the mid-seventeenth century achieved a reunification of the cosmos, society, and the self in terms of a new metaphor — the machine. Developed by the French thinkers Mersenne, Gassendi, and Descartes in the 1620s and 1630s and elaborated by a group of English emigrés to Paris in the 1640s and 1650s, the new mechanical theories emphasized and reinforced elements in human experience developing slowly since the late Middle Ages, but accelerating in the sixteenth century.

New forms of order and power provided a remedy for the disorder perceived to be spreading throughout culture. In the organic world, order meant the function of each part within the larger whole, as determined by its nature, while power was diffused from the top downward through the social or cosmic hierarchies. In the mechanical world, order was redefined to mean the predictable behavior of each part within a rationally determined system of laws, while power derived from active and immediate intervention in a secularized world. Order and power together constituted control. Rational control over nature, society, and the self was achieved by redefining reality itself through the new machine metaphor.

As the unifying model for science and society, the machine has permeated and reconstructed human consciousness so totally that today we scarcely question its validity. Nature, society, and the human body are composed of interchangeable atomized parts that can be repaired or replaced from outside. The "technological fix" mends an ecological malfunction, new human beings replace the old to maintain the smooth functioning of industry and bureaucracy, and interventionist medicine exchanges a fresh heart for a worn-out, diseased one. . . .

The removal of animistic, organic assumptions about the cosmos constituted the death of nature — the most far-reaching effect of the Scientific Revolution. Because nature was now viewed as a system of dead, inert particles moved by external, rather than inherent forces, the mechanical framework itself could legitimate the manipulation of nature. Moreover, as a conceptual framework, the mechanical order had associated with it a framework of values based on power, fully compatible with the directions taken by commercial capitalism . . . .

The mechanistic view of nature, developed by the seventeenth-century natural philosophers and based on a Western mathematical tradition going back to Plato, is still dominant in science today. This view assumes that nature can be divided into parts and that the parts can be rearranged to create other species of being. "Facts" or information bits can be extracted from the environmental context and rearranged according to a set of rules based on logical and mathematical operations. The results can then be tested and verified by resubmitting them to nature, the ultimate judge of their validity. Mathematical formalism provides the criterion for rationality and certainty, nature the criterion for empirical validity and acceptance or rejection of the theory.

The work of historians and philosophers of science notwithstanding, it is widely assumed by the scientific community that modern science is objective, value-free, and context-free knowledge of the external world. To the extent to which the sciences can be reduced to this mechanistic mathematical model, the more legitimate they become as sciences. Thus the reductionist hierarchy of the validity of the sciences first proposed in the nineteenth century by French positivist philosopher August Comte is still widely assumed by intellectuals, the most mathematical and highly theoretical sciences occupying the most revered position.

The mechanistic approach to nature is as fundamental to the twentieth-century revolution in physics as it was to classical Newtonian science, culminating in the nineteenth-century unification of mechanics, thermodynamics, and electromagnetic theory. Twentieth-century physics still views the world in terms of fundamental particles — electrons, protons, neutrons, mesons, muons, pions, taus, thetas, sigmas, pis, and so on. The search for the ultimate unifying particle, the quark, continues to engage the efforts of the best theoretical physicists.

Mathematical formalism isolates the elements of a given quantum mechanical problem, places them in a latticelike matrix, and rearranges them through a mathematical function called an *operator*. Systems theory extracts possibly relevant information bits from the environmental context and stores them in a computer memory for later use. But since it cannot store an infinite number of "facts", it must select a finite number of potentially relevant pieces of data according to a theory or set of rules governing the selection process. For any given solution, this mechanistic approach very likely excludes some potentially relevant factors.

Systems theorists claim for themselves a holistic outlook, because they believe that they are taking into account the ways in which all the parts in a given system affect the whole. Yet the formalism of the calculus of probabilities excludes the possibility of mathematizing the gestalt — that is, the ways in which each part at any given instant take their meaning from the whole. The more open, adaptive, organic, and complex the system, the less successful is the formalism. It is most successful when applied to closed, artificial, precisely defined, relatively simple systems. Mechanistic assumptions about nature push us increasingly in the direction of artificial environments, mechanized control over more and more aspects of human life, and a loss of the quality of life itself.

In the social sphere, the mechanistic model helps to guide technological and industrial development. In *The Technological Society*, Jacques Ellul discussed the techniques of economics and the mechanistic organization of specialties inherent in and entailed by the machines and mathematical methods themselves. The calculating machine, punch card machine, microfilm, and computer transform statistical methods and administrative organization into specialized agencies centered around one or more statistical categories.

Econometric models and stochastics are used to operate on

statistical data in order to analyze, compare, and predict. In social applications, attempts to predict public reaction through the calculus of probabilities may make a public informed of its confirmation to a trend act in the inverse manner.

But the public, by so reacting falls under the influence of a new prediction which is completely determinable . . . . It must be assumed, however, that one remains within the framework of rational behavior. The system works all the better when it deals with people who are better integrated into the mass . . . whose consciousness is partially paralyzed, who lend themselves willingly to statistical observations and systematization.

Such attempts to reduce human behavior to statistical probabilities and to condition it by such psychological techniques as those developed by B. F. Skinner are manifestations of the pervasiveness of the mechanistic mode of thought developed by the seventeenth century scientists. . . .

The March 1979 accident at the Three-Mile Island nuclear reactor near Harrisburg, Pennsylvania, epitomized the problems of the "death of nature" that have become apparent since the Scientific Revolution. The manipulation of nuclear processes in an effort to control and harness nature through technology backfired into disaster. The long-range economic interests and public image of the power company and the reactor's designer were set above the immediate safety of the people and the health of the earth. The hidden effects of radioactive emissions, which by concentrating in the food chain could lead to an increase in cancers over the next several years, were initially downplayed by those charged with responsibility for regulating atomic power.

Three-Mile Island is a recent symbol of the earth's sickness caused by radioactive wastes, pesticides, plastics, photochemical smog, and fluorocarbons. The pollution "of her purest streams" has been supported since the Scientific Revolution by an ideology of "power over nature", an ontology of interchangeable atomic and human parts, and a methodology of "penetration" into her innermost secrets. The sick earth, "yea dead, yea putrified", can probably in the long run be restored to health only by a reversal of mainstream values and a revolution in economic priorities. In this sense, the world must once again be turned upside down.

As natural resources and energy supplies diminish in the future, it will become essential to examine alternatives of all kinds so that, by adopting new social styles, the quality of the environment can be sustained. Decentralization, nonhierarchical forms of organization, recycling of wastes, simpler living styles involving less-polluting "soft" technologies, and labor-intensive rather than capital-intensive economic methods are possibilities only beginning to be explored. The future distribution of energy and resources among communities should be based on the integration of human and natural ecosystems. Such a restructuring of priorities may be crucial if people and nature are to survive.

The Technological Society

JACQUES ELLUL

Although the individual existing in the framework of a civilization of a certain type was always confronted with certain techniques, {he}* was nevertheless free to break with that civilization and to control {his} own individual destiny. The constraints to which {he} was subject did not function decisively. because they were of a non-technical nature and could be broken through. In an active civilization, even one with a fairly good technical development, the individual could always break away and lead, say, a mystical and contemplative life. The fact that techniques and {man} were more or less on the same level permitted the individual to repudiate techniques and get along without them. Choice was a real possibility for {him} not only with regard to {his} inner life, but with regard to the outer form of {his} life as well. The essential elements of life were safeguarded and provided for, more or less liberally, by the very civilization whose forms {he} rejected. In the Roman Empire (a technical civilization in a good many respects), it was possible for a {man} to withdraw and live as a hermit or in the country, apart from the evolution and the principal technical power of the Empire. Roman law was powerless in the face of an individual's decision to evade military service or, to a very great degree, imperial taxes and jurisdiction. Even greater was the possibility of the individual's freedom with respect to material techniques.

There was reserved for the individual an area of free choice at the cost of minimal effort. The choice involved a conscious decision and was possible only because the material burden of technique had not yet become more than a {man} could shoulder. The existence of choice, a result of characteristics we have already discussed, appears to have been one of the most important historical factors governing technical evolution and revolution. Evolution was not, then, a logic of discovery or an inevitable progression of techniques. It was an interaction of technical effectiveness and effective human decision. Whenever either one of these elements disappeared, social and human stagnation necessarily followed. Such was the case, for example, when effective technique was (or became) rudimentary

41

* See "Questioning Technology, Questioning Patriarchy," p. 1.

and inefficacious among the Negroes of Africa. As to the consequences of a lapse in the second element, we are experiencing them today.

The New Characteristics

The characteristics of the relationship of technique, society, and the individual which we have analyzed were, I believe, common to all civilizations up to the eighteenth century. Historically, their existence admits of little discussion. Today, however, the most cursory review enables us to conclude that all these characteristics have disappeared. The relation is not the same; it does not present any of the constants recognizable until now. But that is not sufficient to characterize the technical phenomenon of our own day. This description would situate it in a purely negative perspective, whereas the technical phenomenon is a positive thing; it presents positive characteristics which are peculiar to it. The old characteristics of technique have indeed disappeared; but new ones have taken their place. Today's technical phenomenon, consequently, has almost nothing in common with the technical phenomenon of the past. I shall not insist on demonstrating the negative aspect of the case, the disappearance of the traditional characteristics. To do so would be artificial, didactic, and difficult to defend. I shall point out, then, in a summary fashion, that in our civilization technique is in no way limited. It has been extended to all spheres and encompasses every activity, including human activities. It has led to a multiplication of means without limit. It has perfected indefinitely the instruments available to man and put at his disposal an almost limitless variety of intermediaries and auxiliaries. Technique has been extended geographically so that it covers the whole earth. It is evolving with a rapidity disconcerting not only to the man in the street but to the technician himself. It poses problems which recur endlessly and ever more acutely in human social groups. Moreover, technique has become objective and is transmitted like a physical thing; it leads thereby to a certain unity of civilization, regardless of the environment or the country in which it operates. We are faced with the exact opposite of the traits previously in force. We must, therefore, examine carefully the positive characteristics of the technique of the present.

There are two essential characteristics of today's technical phenomenon which I shall not belabor because of their obviousness.

These two, incidentally, are the only ones which, in general, are emphasized by the "best authors".

The first of these obvious characteristics is rationality. In technique, whatever its aspect or the domain in which it is applied, a rational process is present which tends to bring mechanics to bear on all that is spontaneous or irrational. This rationality, best exemplified in systematization, division of labor, creation of standards, production norms, and the like, involves two distinct phases: first, the use of "discourse" in every operation; this excludes spontaneity and personal creativity. Second, there is the reduction of method to its logical dimension alone. Every intervention of technique is, in effect, a reduction of facts, forces, phenomena, means, and instruments to the schema of logic.

The second obvious characteristic of the technical phenomenon is artificiality. Technique is opposed to nature. Art, artifice, artificial: technique as art is the creation of an artificial system. This is not a matter of opinion. The means{man}has at his disposal as a function of technique are artificial means. For this reason, the comparison proposed by Emmanuel Mounier between the machine and the human body is valueless. The world that is being created by the accumulation of technical means is an artificial world and hence radically different from the natural world.

It destroys, eliminates, or subordinates the natural world, and does not allow this world to restore itself or even to enter into a symbiotic relation with it. The two worlds obey different imperatives, different directives, and different laws which have nothing in common. Just as hydroelectric installations take waterfalls and lead them into conduits, so the technical milieu absorbs the natural. We are rapidly approaching the time when there will be no longer any natural environment at all. When we succeed in producing artificial *aurorae boreales*, night will disappear and perpetual day will reign over the planet . . . .

There is no personal choice, in respect to magnitude, between, say, 3 and 4; 4 is greater than 3; this is a fact which has no personal reference. No one can change it or assert the contrary or personally escape it. Similarly, there is no choice between two technical methods. One of them asserts itself inescapably: its results are calculated, measured, obvious, and indisputable.

A surgical operation which was formerly not feasible but can now be performed is not an object of choice. It simply is. Here we see the prime aspect of technical automatism. Technique itself, *ipso facto* and without indulgence or possible discussion, selects among the means to be employed. The human being is no longer in any sense the agent of choice. Let no one say that man is the agent of technical progress (a question I shall discuss later) and that it is he who chooses among possible techniques. In reality, he neither is nor does anything of the sort. He is a device for recording effects and results obtained by various techniques. He does not make a choice of complex and, in some way, human motives. He can decide only in favor of the technique that gives the maximum efficiency. But this is not choice. A machine could effect the same operation. Man still appears to be choosing when he abandons a given method that has proved excellent from some point of view. But his action comes solely from the fact that he has thoroughly analyzed the results and determined that from another point of view the method in question is less efficient. A good example is furnished by the attempts to deconcentrate our great industrial plants after we had concentrated them to the maximum possible degree. Another example would be the decision to abandon certain systems of high production in order to obtain a more constant productivity, although it might be less per capita. It is always a question of the improvement of the method in itself.

The worst reproach modern society can level is the charge that some person or system is impeding this technical automatism. When a labor union leader says: "In a period of recession, productivity is a social scourge", his declaration stirs up a storm of protest and condemnation, because he is putting a personal judgement before the technical axiom that what can be produced must be produced. If a machine can yield a given result, it must be used to capacity, and it is considered criminal and antisocial not to do so. Technical automatism may not be judged or questioned; immediate use must be found for the most recent, efficient, and technical process.

Communism's fundamental criticism of capitalism is that financial capitalism checks technical progress that produces no profits; or that it promotes technical progress only in order to reserve for itself a monopoly. In any case, as Rubinstein points out, technical progress occurs under capitalism for reasons which have nothing to do with technique, and it is this fact which is to be criticized. Since the Communist regime is oriented toward technical progress, the mark of the superiority of Communism is that it adopts all technical progress. Rubinstein concludes his study by remarking that this progress is the goal of all efforts in the Soviet Union, where it is said to be possible to allow free play to technical automatism without checking it in any way.

Another traditional analysis supplements Rubinstein's. This serious study, carried out by Thorstein Veblen, maintains that there is a conflict between the machine and business. Financial investment, which originally accelerated invention, now prolongs technical inactivity. Capitalism does not give free play to technical activity, the goal of which is that a more efficient method or a more rapidly acting machine should *ipso facto* and automatically replace the preceding method or machine. Capitalism does not give free play to these factors because it inadmissibly subordinates technique to ends other than technique itself, and because it is incapable of absorbing technical progress. The replacement of machines at the tempo of technical invention is completely impossible for capitalist enterprise because there is no time to amortize one machine before new ones appear. Moreover, the more these machines are improved, and hence become more efficient, the more they cost.

The pursuit of technical automatism would condemn capitalist enterprises to failure. The reaction of capitalism is well known: the patents of new machines are acquired and the machines are never put into operation. Sometimes machines that are already in operation are acquired, as in the case of England's largest glass factory in 1932, and destroyed. Capitalism is no longer in a position to pursue technical automatism on the economic or social plane.

3. How has industrial technology adversely affected individuals, societies, and the planet as a whole?

The anthropologist Gene Weltfish once told of interviewing a very old Pawnee woman who had watched emigrant wagon trains roll westward across the plains. "We told each other that for them to bear the suffering of their journey," she told Weltfish, "it must have been really awful where they came from."

Not the usual perspective on pioneer-Indian relationships! And with their arrival in great numbers in the West, Euroamerican settlers put an abrupt end to stable hunter-gatherer cultures that had persisted for thousands of years, as these settlers proceeded to recreate what they had left behind.

"Pioneers" are still hacking away at the last remaining "frontiers", in the Amazon Basin, the Arctic, and the Antarctic. Wilderness, the reservoir of Earth's nutrients, has become quaint and confined, as small in proportion to the planet as parks are in proportion to the commercial and residential city.

Our awareness of the precariousness of this state comes and goes. Awareness brings us to despair, while denial robs us of our affiliation with the natural world. Are the positive benefits of medical technology and domestic comfort worth the cost in almost every other area of human life, not to mention all life on earth?

George Bradford answers with an anguished summation of the political economy of technology out of control. Siegel and Markoff have studied the new high-tech mecca, Silicon Valley, and found appalling, not promising, prospects, while Morris Berman proposes that modernity itself must be deeply reordered if we are to deal with the new world wrought by industrial technology.

We All Live in Bhopal

GEORGE BRADFORD

The cinders of the funeral pyres at Bhopal are still warm, and the mass graves still fresh, but the media prostitutes of the corporations have already begun their homilies in defense of industrialism and its uncounted horrors. Some 3,000 people were slaughtered in the wake of the deadly gas cloud, and 20,000 will remain permanently disabled. The poison gas left a 25 square mile swath of dead and dying, people and animals, as it drifted southeast away from the Union Carbide factory. "We thought it was a plague", said one victim. Indeed it was: a chemical plague, an *industrial plague,*
Ashes, ashes, all fall down!

A terrible, unfortunate, "accident", we are reassured by the propaganda apparatus for Progress, for History, for "Our Modern Way of Life". A price, of course, has to be paid — since the risks are necessary to ensure a higher Standard of Living, a Better Way of Life.

The *Wall Street Journal*, tribune of the bourgeoisie, editorialized, "It is worthwhile to remember that the Union Carbide insecticide plant and the people surrounding it were where they were for compelling reasons. India's agriculture has been thriving, bringing a better life to millions of rural people, and partly because of the use of modern agricultural technology that includes applications of insect killers." The indisputable fact of life, according to this sermon, is that universal recognition that India, like everyone else, "needs technology. Calcutta-style scenes of human deprivation can be replaced as fast as the country imports the benefits of the West's industrial revolution and market economics." So, despite whatever dangers are involved, "the benefits outweigh the costs". (Dec. 13 1984)

The *Journal* was certainly right in one regard — the reasons for the plant and the people's presence there are certainly compelling:

47

capitalist market relations and technological invasion are as compelling as a hurricane to the small communities from which those people were uprooted. It conveniently failed to note, however, that countries like India do not import the *benefits* of industrial capitalism; those benefits are *exported* in the form of loan repayments to fill the coffers of the bankers and corporate vampires who read the *Wall Street Journal* for the latest news of their investments. The Indians only take the risks and pay the costs; in fact, for them, as for the immiserated masses of people living in the shantytowns of the Third World, there are no risks, only certain hunger and disease, only the certainty of death squad revenge for criticizing the state of things as they are.

Green Revolution a Nightmare

In fact, the Calcutta-style misery is the result of Third World industrialization and the so-called industrial "Green Revolution" in agriculture. The Green Revolution, which was to revolutionize agriculture in the "backward" countries and produce greater crop yields, has only been a miracle for the banks, corporations and military dictatorships who defend them. The influx of fertilizers, technology, insecticides and bureaucratic administration exploded millennia-old rural economies based on subsistence farming, creating a class of wealthier farmers dependent upon western technologies to produce cash crops such as coffee, cotton and wheat for export, while the vast majority of farming communities were destroyed by capitalist market competition and sent like refugees into the growing cities. These victims, paralleling the destroyed peasantry of Europe's Industrial Revolution several hundred years before, joined either the permanent underclass of unemployed and underemployed slumdwellers eking out a survival on the tenuous margins of civilization, or became proletarian fodder in the Bhopals, Sao Paulos and Djakartas of an industrializing world — an industrialization process, like all industrialization in history, paid for by the pillage of nature and human beings in the countryside.

Food production goes up in some cases, of course, because the measure is only quantitative — some foods disappear while others are produced year round, even for export. *But subsistence is destroyed.* Not only does the rural landscape begin to suffer the consequences of constant crop production and use of chemicals, but the masses of people — laborers on the land and in the teeming hovels growing around the industrial plants — go hungrier in a vicious cycle of exploitation, while the wheat goes abroad to buy absurd commodities and weapons.

But subsistence is culture as well: culture is destroyed with subsistence, and people are further trapped in the technological labyrinth. The ideology of progress is there, blared louder than ever by those with something to hide, a cover-up for plunder and murder on levels never before witnessed.

Industrialization of the Third World

The industrialization of the Third World is a story familiar to anyone who takes even a glance at what is occurring. The colonial countries are nothing but a dumping ground and pool of cheap labor for capitalist corporations. Obsolete technology is shipped there along with the production of chemicals, medicines and other products banned in the developed world. Labor is cheap, there are few if any safety standards, and *costs are cut*. But the formula of cost-benefit still stands: the costs are simply borne by others, by the victims of Union Carbide, Dow, and Standard Oil.

Chemicals found to be dangerous and banned in the US and Europe are produced instead overseas — DDT is a well-known example of an enormous number of such products, such as the unregistered pesticide Leptophos exported by the Velsicol Corporation to Egypt which killed and injured many Egyptian farmers in the mid-1970's. Other products are simply dumped on Third World markets, like the mercury-tainted wheat which led to the deaths of as many as 5,000 Iraqis in 1972, wheat which had been imported from the US. Another example was the wanton contamination of Nicaragua's Lake Managua by a chlorine and caustic soda factory owned by Pennwalt Corporation and other investors, which caused a major outbreak of mercury poisoning in a primary source of fish for the people living in Managua.

Union Carbide's plant at Bhopal did not even meet US safety standards according to its own safety inspector, but a UN expert on international corporate behavior told the *New York Times*, "A whole list of factors is not in place to insure adequate industrial safety" throughout the Third World. "Carbide is not very different from any other chemical company in this regard." According to the *Times*, "In a Union Carbide battery plant in Jakarta, Indonesia, more than half the workers had kidney damage from mercury exposure. In an asbestos cement factory owned by the Manville Corporation 200 miles west of Bhopal, workers in 1981 were routinely covered with

asbestos dust, a practice that would never be tolerated here."
(12/9/84)

Some 22,500 people are killed every year by exposure to
insecticides — a much higher percentage of them in the Third
World than use of such chemicals would suggest. Many experts
decried the lack of an "industrial culture" in the "underdeveloped"
countries as a major cause of accidents and contamination. But
where an "industrial culture" thrives, is the situation really much
better?

Industrial Culture and Industrial Plague

In the advanced industrial nations an "industrial culture" (and
little other) exists. Have such disasters been avoided as the claims of
these experts would lead us to believe?

Another event of such mammoth proportions as those of Bhopal
would suggest otherwise — in that case, industrial pollution killed
some 4,000 people in a large population center. That was London, in
1952, when several days of "normal" pollution accumulated in
stagnant air to kill and permanently injure thousands of Britons.

Then there are the disasters closer to home or to memory, for
example, the Love Canal (still leaking into the Great Lakes water
system), or the massive dioxin contaminations at Seveso, Italy and
Times Creek, Missouri, where thousands of residents had to be
permanently evacuated. And there is the Berlin and Farro dump at
Swartz Creek, Michigan, where C-56 (a pesticide by-product of
Love Canal fame), hydrochloric acid and cyanide from Flint auto
plants had accumulated. "They think we're not scientists and not
even educated", said one enraged resident, "but anyone who's been
in high school knows that cyanide and hydrochloric acid is what
they mixed to kill the people in the concentration camps".

A powerful image: industrial civilization as one vast, stinking
extermination camp. We all live in Bhopal, some closer to the gas
chambers and to the mass graves, but all of us close enough to be
victims. And Union Carbide is obviously not a fluke — the poisons
are vented in the air and water, dumped in rivers, ponds and
streams, fed to animals going to market, sprayed on lawns and
roadways, sprayed on food crops, every day, everywhere. The result
may not be as dramatic as Bhopal (which then almost comes to
serve as a *diversion*, a deterrence machine to take our minds off the
pervasive reality which Bhopal truly represents), but it is as deadly.
When ABC News asked University of Chicago professor of public

health and author of *The Politics of Cancer*, Jason Epstein, if he
thought a Bhopal-style disaster could occur in the US, he replied: "I
think what we're seeing in America is far more slow — not such
large accidental occurrences, but a slow, gradual leakage with the
result that you have excess cancers or reproductive abnormalities."

In fact, birth defects have doubled in the last 25 years. And cancer
is on the rise. In an interview with the *Guardian*, Hunter College
professor David Kotelchuck described the "Cancer Atlas" maps
published in 1975 by the Department of Health, Education and
Welfare. "Show me a red spot on these maps and I'll show you an
industrial center of the US", he said. "There aren't any place names
on the maps but you can easily pick out concentrations of industry.
See, it's not Pennsylvania that's red it's just Philadelphia, Erie and
Pittsburgh. Look at West Virginia here, there's only two red spots,
the Kanawha Valley, where there are nine chemical plants
including Union Carbide's, and this industrialized stretch of the
Ohio River. It's the same story wherever you look."

There are 50,000 toxic waste dumps in the United States. The
EPA admits that *ninety per cent* of the 90 billion pounds of toxic waste
produced annually by US industry (70 per cent of it by chemical
companies) is disposed of "improperly" (although we wonder what
they would consider "proper" disposal). These deadly products of
industrial civilization — arsenic, mercury, dioxin, cyanide, and
many others — are simply dumped, "legally" and "illegally",
wherever convenient to industry. Some 66,000 different compounds
are used in industry. Nearly a billion tons of pesticides and
herbicides comprising 225 different chemicals were produced in the
US last year, and an additional 79 million pounds were imported.
Some two per cent of chemical compounds have been tested for side
effects. There are 15,000 chemical plants in the United States, daily
manufacturing mass death.

All of the dumped chemicals are leaching into our water. Some
three to four thousand wells, depending on which government
agency you ask, are contaminated or closed in the US. In Michigan
alone, 24 municipal water systems have been contaminated, and a
thousand sites have suffered major contamination. According to the
Detroit *Free Press*, "The final toll could be as many as 10,000 sites" in
Michigan's "water wonderland" alone (April 15, 1984).

And the coverups go unabated here as in the Third World. One
example is that of dioxin; during the proceedings around the Agent
Orange investigations, it came out that Dow Chemical had lied all

along about the effects of dioxin. Despite research findings that dioxin is "exceptionally toxic" with "a tremendous potential for producing chlor-acne and systemic injury", Dow's top toxicologist, V. K. Rowe, wrote in 1965, "We are not in any way attempting to hide our problems under a heap of sand. But we certainly do not want to have any situations arise which will cause the regulatory agencies to become restrictive."

Now Vietnam suffers a liver cancer epidemic and a host of cancers and health problems caused by the massive use of Agent Orange there during the genocidal war waged by the US. The sufferings of the US veterans are only a drop in the bucket. And dioxin is appearing everywhere in our environment as well, in the form of recently discovered "dioxin rain".

Going to the Village

When the Indian authorities and Union Carbide began to process the remaining gases in the Bhopal plant, thousands of residents fled, despite the reassurances of the authorities. The *New York Times* quoted one old man who said, "They are not believing the scientists or the state government or anybody. They only want to save their lives."

The same reporter wrote that one man had gone to the train station with his goats, "hoping that he could take them with him — anywhere, as long as it was away from Bhopal" (Dec. 14, 1984). The same old man quoted above told the reporter, "All the public has gone to the village." The reporter explained that "going to the village" is what Indians do when trouble comes.

A wise and age-old strategy for survival by which little communities always renewed themselves when bronze, iron and golden empires with clay feet fell to their ruin. But subsistence has been and is everywhere being destroyed, and with it, culture. What are we to do when there is no village to go to? When we all live in Bhopal, and Bhopal is everywhere? The comments of two women, one a refugee from Times Creek, Missouri, and another from Bhopal, come to mind. The first woman said of her former home, "This was a nice place once. Now we have to bury it." The other woman said, "Life cannot come back. Can the government pay for the lives? Can you bring those people back?"

The corporate vampires are guilty of greed, plunder, murder, slavery, extermination and devastation. And we should avoid any pang of sentimentalism when the time comes for them to pay for their crimes against humanity and the natural world. But we will have to go beyond them, to ourselves: subsistence, and with it culture, has been destroyed. We have to find our way back to the village, out of industrial civilization, out of this exterminist system.

The Union Carbides, the Warren Andersons, the "optimistic experts" and the lying propagandists all must go, but with them must go the pesticides, the herbicides, the chemical factories and the chemical way of life which is nothing but death.

Because this is Bhopal, and it is all we've got. This "once nice place" can't be simply buried for us to move on to another pristine beginning. The empire is collapsing. We must find our way back to the village, or as the North American natives said, "back to the blanket", and we must do this not by trying to save an industrial civilization which is doomed, but in that renewal of life which must take place in its ruin. By throwing off this Modern Way of Life, we won't be "giving things up" or sacrificing, but throwing off a terrible burden. Let us do so soon before we are crushed by it.

The High Cost of High Tech: The Dark Side of the Chip

LENNY SIEGEL & JOHN MARKOFF

The belt of industrial communities at the southern edge of the San Francisco Bay universally symbolizes the promise of the microelectronics era. It was first called Silicon Valley in the early 1970s, when manufacturers of silicon chips became the Santa Clara Valley's major employers. The Valley is home to the greatest concentration of high-tech professionals and enterprises in the world. It is a land where the information-rich, particularly those trained in science and technology, can make both their mark and their millions.

Though Silicon Valley is in many ways unique, planners, officials, and commercial interests throughout the country see the area as a model for industrial growth in the information age. While few other areas can hope to rival the Valley, many have already attracted their share of high-tech facilities. As high tech grows, they will learn the harsh truth behind the legends of Silicon Valley.

Many of the Valley's problems are directly caused by high tech. Others are found elsewhere, but they are significant merely because the residents of would-be Silicon Valleys have been told that the electronics industry has no serious problems. If they study the lessons of the Valley, they can avoid many of the pitfalls of high-tech growth.

"Maria", a 26-year-old political refugee from Argentina, found work in Silicon Valley, but she did not strike gold. She quit her $4.10 an hour production job at Memorex to have her first baby. For two years, she illegally stuffed and soldered thousands of printed circuit (PC) boards in her home. Her employer, a middle-aged woman she calls "Lady", sub-contracted assembly work from big firms — so Maria was told — like Apple and Memorex.

Maria gladly accepted the low piece-rate work because child care would have eaten up most of her after-tax earnings at a full-time job. She quit, however, when Lady asked her to wash her assembled boards by dipping them into a panful of solvent, heated on her kitchen stove. Maria, unlike most Silicon Valley cottage workers, had studied chemistry before immigrating into the US and she knew that the hydrocarbon fumes could make her young son, crawling around on the kitchen floor, seriously ill.

Lady contracts with about a hundred minority women, primarily immigrants and refugees from Latin America, Korea, and Indochina. Although semiconductor chips are fabricated with precise machinery in super-clean rooms, they can be attached by hand, anywhere, to the printed circuit boards that form the heart of most computer equipment.

Silicon Valley's workforce is sharply stratified. In the electronics industry, pay, status, and responsibility are primarily a function of education. The professionals who make the Valley unique sit at the top of the occupational ladder; they are paid well, and the ambitious among them can make millions. Most are white men, but Japanese-Americans and ethnic Chinese are over-represented as well.

The world of Silicon Valley's managers and professionals is centered in northern Santa Clara County, near Stanford University and the historical center of the Valley's high-tech industry. Unlike the white-collar workers who commute to America's established downtown areas, Silicon Valley's affluent have chosen to live near their place of work. Other new, high-tech centers appear to be developing along a remarkably similar pattern.

Since Stanford University established its Industrial Park in 1951, high-tech companies have clustered near the university. The Industrial Park, on Stanford-owned land just a mile from the academic campus, established standards for industrial development in Silicon Valley, and it is still considered a model throughout North America. For three decades, its low-slung buildings, innovative architecture, and expanses of green landscape perpetuated the belief that high tech was a clean industry and a good neighbor. The suburbs around Stanford have long been known for their attractive living environment and good schools; and commuting, even before the 1973 rise in oil prices, was uncomfortable, costly, and time-consuming. So professional workers generally bought homes or rented as close to work as possible.

As the Valley boomed, its industrial core spread, but until the 1980s this core was for the most part confined to the northern, suburban portion. Like their predecessors, the engineers, scientists, and managers who came to the Valley from all over the world settled near their jobs. This influx of high-income families drove up the cost of housing. By the 1970s, rents and prices in the Valley were among the highest in the nation.

By and large, the unemployed, the service workers, and the Valley's low-paid production workers — who have always earned a fraction of the professionals' salaries — were driven from the centers of employment. San Jose, the county's traditional urban center and home to half its residents, became a bedroom community for the production workforce.

Palo Alto, which receives property and sales tax revenues from the Stanford Industrial Park, easily provides municipal services to its relatively affluent citizens. San Jose, on the other hand, has a much smaller tax base from which it must serve the county's poorer residents. Production workers from San Jose spend their days in the north county, generating wealth for electronics companies to pay into suburban treasuries. They then return to homes protected by San Jose's underfunded police and fire departments and streets poorly maintained by its public works department.

Nowhere are the two worlds of Silicon Valley further apart than in education. Palo Alto's public school system is considered among the best in the nation. In fact, that is a major reason why high-tech professionals move to the area. In 1983, however, the San Jose Unified School District, the largest of several districts in the city, became the first American school system since 1943 to declare bankruptcy.

A few years back, several women on the morning shift at Verbatim, a Silicon Valley manufacturer of memory disks for computers, complained of dizziness, shortness of breath, and weakness. Some even reported seeing a haze in the factory air. More than 100 people were quickly evacuated from the building, and the company sent 35 of them to a nearby industrial clinic.

Hours later, inspectors from the California Occupational Safety and Health Administration could not find fumes intense enough to explain the complaints, and they termed the episode "mass psychogenic illness", also known as assembly-line hysteria. In the stressful world of high-volume electronics assembly, mass hysteria is not unknown. But chances are high that the Verbatim workers' bodies had detected the presence of toxic chemicals at a level below the threshhold recognized by health officials.

High-tech industry's environmentally controlled "clean rooms", in which electronics workers must wear surgical gowns and gloves, are not designed to protect the workers; they are built to protect microelectronic products against particulate contamination. Despite the protective clothing, equipment, and vents found at a typical semiconductor plant, in the pressure to meet production quotas many Silicon Valley workers are frequently exposed to hazardous liquids and fumes.

The hazardous materials used in semi-conductor production include acids, cyanide compounds, organic solvents, and silicon tetrachloride, which turns into hydrochloric acid when its fumes are inhaled into the lungs. Arsine gas, a lethal form of arsenic, can cause serious damage to the liver, heart, and blood cells, even when inhaled in small quantities. It has been used extensively for years in the production of silicon chips. Now, as the Pentagon is promoting the development and production of chips based upon gallium arsenide instead of silicon, the likelihood of workers being exposed to arsenic is growing.

It is possible that communities and regions which study the lessons of Silicon Valley can substantially reduce the risk high-tech production poses to the environment and public health. Unfortunately, high tech's environmental record has not leaked out to the rest of the country. Officials who promote high tech as a solution to local or regional economic ills paint a picture of the industry as shiny as the surface of a silicon wafer. They call high tech a "sunrise industry", clean and light in contrast to "smokestack" industries like steel and auto production, known for their drab, monstrous factories and ever present plumes of vapor and smoke.

It isn't hard to see where high tech got its reputation. Electronic products — chips, computers, switchboards, and so on — don't breathe exhaust or drip oil. The factories are rambling, well-landscaped buildings, resembling modern college libraries; no smokestacks protrude above their facades. Many production steps take place in so-called clean rooms, where the air is fanatically filtered and production workers wear surgical gowns. But the industry's vast investment in cleanliness is designed principally to protect microelectronic components from the dust particles that could prevent them from functioning properly. It does not protect high-tech's workers nor the residents who live in the communities that surround the plants, from the toxic chemicals and metals essential to high-tech manufacturing.

One of the greatest ironies of micro-electronics technology is that the transformation of America into an information society relies, at its core, upon a technology from the industrial era: chemical processing. The manufacture of chips, printed circuit boards, magnetic media, and other high-tech products uses some of the most dangerous materials known to humanity. And the accidental release of those toxins into the air, the ground, and bodies of water poses a significant threat to public health.

High-tech pollution is a fact of life wherever the industry has operated for any length of time, from Malaysia to Massachusetts. Yet nowhere has the growing threat that electronics production poses to public health been clearer than in Silicon Valley, where the concentration of high-tech production has greatly magnified the industry's environmental problems.

The hazards of high tech have become increasingly clear during the past few years, but it may be decades before the full impact on public health is known. The electronics industry uses thousands of different toxic materials, yet the volume is small compared to chemical-intensive industries such as petroleum and pesticide production. Still, a Bhopal-like incident, in which hundreds of people are killed immediately from a single leak, is a serious possibility.

Even without such a catastrophic accident, however, the long-term toll from high-tech pollution may be enormous. High-tech toxics have been slowly entering the environment of Silicon Valley for decades. Though widely used chemicals such as hydrocarbon solvents are known to cause ailments ranging from headaches and birth defects to cancer, it is difficult to demonstrate that any particular person is a victim of a particular leak or spill. But there is no doubt that industrial chemicals are affecting the health of growing numbers of people.

San Jose attorney Amanda Hawes is one of a handful of Silicon Valley activists who warned for years that high tech was indeed a hazardous industry. She has built up her reputation by representing electronics workers injured by chemicals on the job. Today she also represents residents of the Los Paseos neighborhood in southern San Jose. A new, comfortable, working-class suburb typical of Silicon Valley, Los Paseos is distinguished by the presence of a chip manufacturing factory built by Fairchild Semiconductor in 1975.

Hawes carries with her a large zoning map of the area surrounding the Fairchild plant. On every block in the surrounding neighborhood there are several colored pins and flags. Each triangular red flag represents a child born with heart anomalies; each blue pin marks a miscarriage; each yellow flag signals a cancer case. Black flags, superimposed on the other markers, note recent deaths. Hawes also carries with her a supply of pins, and she must frequently add one to the display. She charges that Fairchild is responsible for the area's high incidence of disease.

Most of Hawes's clients believed that electronics was a pollution-free industry until January 1982. At that time, officials disclosed that six weeks earlier they had shut down a drinking water well operated by the Great Oaks Water Co., just 2,000 feet from an underground storage tank, including suspected carcinogens trichloroethane and dichloroethylene, had entered the water supply. When residents learned of the leak, they quickly concluded that the company was to blame for the area's alarmingly high incidence of birth defects and miscarriages.

Since then, Fairchild spent at least $15 million to reduce the concentration of solvents in the aquifer, but the water will never be as clean as it was before Fairchild set up shop there. Now the factory stands empty, a monument to the dying myth of high tech as a clean, light industry.

The Fairchild leak exploded onto the local front pages and six o'clock news, breaking through a long-standing barrier of silence on high-tech pollution. The Bay Area press, public officials, and electronics corporations themselves have all been forced to investigate environmental hazards that nobody wanted to believe existed.

Today, scarcely a week passes without the revelation of a new leaking storage tank, poisoned well, or pollution law violation. As soon as the extent of the Fairchild leak was known, other companies started to test the ground water around their underground chemical tanks, and the Bay Area's Regional Water Quality Control Board ordered a comprehensive testing program. Most of the Valley's large production sites were checked — and most came up dirty. Even firms with a reputation for environmental concern, like Hewlett-Packard, had been leaking dangerous toxics used in their manufacturing processes.

Leaks were found at scores of industrial locations within Santa Clara County, but many small facilities have still not been tested.

Nineteen high-tech sites have been placed on the Environmental Protection Agency's "Superfund" list. Nine public and more than sixty private wells have been shut down; many others contain legally allowed levels of contamination. Luckily, Silicon Valley residents have thus far been spared an outright environmental disaster. The Valley's largest source of drinking water is protected by a 200-foot layer of clay, which separates polluted ground water from deep aquifers.

Though Fairchild and nearby IBM began the task of clean-up soon after pollution from their facilities was discovered, many Valley electronics firms have not done much more than sink test wells to determine the extent of their leaks. Pools of hazardous chemicals drift around underground, poisoning shallow private wells and possibly finding a route — for example, via an abandoned agricultural well — to the public water supply. Unless the toxic chemicals are removed or neutralized before they percolate through the clay, the primary water supply of several hundred thousand people will be permanently poisoned. Silicon Valley is sitting on a toxic time bomb. No one knows when it is set to go off; certainly, not enough is being done to defuse it.

The Re-enchantment of the World — I

MORRIS BERMAN

The view of nature which predominated in the West down to the eve of the Scientific Revolution was that of an enchanted world. Rocks, trees, rivers, and clouds were all seen as wondrous, alive, and human beings felt at home in this environment. The cosmos, in short, was a place of *belonging*. A member of this cosmos was not an alienated observer of it but a direct participant in its drama. {His}* personal destiny was bound up with its destiny, and this relationship

*See "Questioning Technology, Questioning Patriarchy," p. 1.

gave meaning to his life. This type of consciousness — what I shall refer to in this book as "participating consciousness" — involves merger, or identification, with one's surroundings, and bespeaks a psychic wholeness that has long since passed from the scene. Alchemy, as it turns out, was the last great coherent expression of participating consciousness in the West.

The story of the modern epoch, at least on the level of mind, is one of progressive disenchantment. From the sixteenth century on, mind has been progressively expunged from the phenomenal world. At least in theory, the reference points for all scientific explanation are matter and motion — what historians of science refer to as the "mechanical philosophy". Developments that have thrown this world view into question — quantum mechanics, for example, or certain types of contemporary ecological research — have not made any significant dent in the dominant mode of thinking. That mode can best be described as disenchantment, nonparticipation, for it insists on a rigid distinction between observer and observed. Scientific consciousness is alienated consciousness: there is no ecstatic merger with nature, but rather total separation from it. Subject and object are always seen in opposition to each other. I am not my experiences, and thus not really a part of the world around me. The logical end point of this world view is a feeling of total reification: everything is an object, alien, not-me; and I am ultimately an object too, an alienated "thing" in a world of other, equally meaningless things. This world is not of my own making; the cosmos cares nothing for me, and I do not really feel a sense of belonging to it. What I feel, in fact, is a sickness in the soul.

Translated into everyday life, what does this disenchantment mean? It means that the modern landscape has become a scenario of "mass administration and blatant violence"[1], a state of affairs now clearly perceived by the man in the street. The alienation and futility that characterized the perceptions of a handful of intellectuals at the beginning of the century have come to characterize the consciousness of the common man at its end. Jobs are stupefying, relationships vapid and transient, the arena of politics absurd. In the vacuum created by the collapse of traditional values, we have hysterical evangelical revivals, mass conversions to the Church of the Reverend Moon, and a general retreat into the oblivion provided by drugs, television and tranquilizers. We also have a desperate search for therapy, by now a national obsession, as millions of Americans try to reconstruct their lives amidst a pervasive feeling of anomie and cultural disintegration. An age in which depression is a norm is a grim one indeed.

Perhaps nothing is more symptomatic of this general malaise than the inability of the industrial economies to provide meaningful work. Some years ago, Herbert Marcuse described the blue- and white-collar classes in America as "one-dimensional". "When technics becomes the universal form of material production", he wrote, "it circumscribes an entire culture; it projects a historical totality — a 'world'" One cannot speak of alienation as such, he went on, because there is no longer a self to be alienated. We have all been bought off, we all sold out to the System long ago and now identify with it completely. "People recognize themselves in their commodities", Marcuse concluded; they have become what they own.[2]

Marcuse's is a plausible thesis. We all know the next-door neighbor who is out there every Sunday, lovingly washing his car with an ardor that is almost sexual. Yet the actual data on the day-to-day life of the middle and working classes tend to refute Marcuse's notion that for these people, self and commodities have merged, producing what he terms the "Happy Consciousness". To take only two examples, Studs Terkel's interviews with hundreds of Americans, drawn from all walks of life, revealed how hollow and meaningless they saw their own vocations. Dragging themselves to work, pushing themselves through the daily tedium of typing, filing, collecting insurance premiums, parking cars, interviewing welfare applicants, and largely fantasizing on the job — these people, says Terkel, are no longer characters out of Charles Dickens, but out of Samuel Beckett.[3] The second study, by Sennett and Cobb, found that Marcuse's notion of the mindless consumer was totally in error. The worker is not buying goods because he identifies with the American Way of Life, but because he has enormous anxiety about his self, which he feels possessions might assuage. Consumerism is paradoxically seen as a way *out* of a system that has damaged him and that he secretly despises; it is a way of trying to keep *free* from the emotional grip of this system.[4]

But keeping free from the System is not a viable option. As technological and bureaucratic modes of thought permeate the deepest recesses of our minds, the preservation of psychic space has become almost impossible.[5] "High-potential candidates" for management positions in American corporations customarily undergo a type of finishing-school education that teaches them how to communicate persuasively, facilitate social interaction, read body language, and so on. This mental framework is then imported into the sphere of personal and sexual relations. One thus learns, for example, how to discard friends who may prove to be career obstacles and to acquire new acquaintances who will assist in one's advancement. The employee's{wife}is also evaluated as an asset or liability in terms of{her}diplomatic skills. And for most males in the industrial nations, the sex act itself has literally become a project, a matter of carrying out the proper techniques so as to achieve the prescribed goal and thus win the desired approval. Pleasure and intimacy are seen almost as a hindrance to the act. But once the ethos of technique and management has permeated the spheres of sexuality and friendship, there is literally no place left to hide. The "widespread climate of anxiety and neurosis" in which we are immersed is thus inevitable . . .[6]

The statistics that reflect this condition in America alone are so grim as to defy comprehension. There is now a significant suicide rate among the seven-to-ten age group, and teenage suicides tripled between 1966 and 1976 to roughly thirty per day. More than half the patients in American mental hospitals are under twenty-one. In 1977, a survey of nine- to eleven-year-olds on the West Coast found that nearly half the children were regular users of alcohol, and that huge numbers in this age group regularly came to school drunk. Dr. Darold Treffert, of Wisconsin's Mental Health Institute, observed that millions of children and young adults are now plagued by "a gnawing emptiness or meaninglessness expressed not as a fear of what may happen to them, but rather as a fear that nothing will happen to them". Official figures from government reports released during 1971-72 recorded that the United States has 4 million schizophrenics, 4 million seriously disturbed children, 9 million alcoholics, and 10 million people suffering from severely disabling depression. In the early 1970s, it was reported that 25 million adults were using Valium; by 1980, Food and Drug Administration figures indicated that Americans were downing benzodiazepines (the class of tranquilizers which includes Valium) at a rate of 5 billion pills a year. Hundreds of thousands of the nation's children, according to *The Myth of the Hyperactive Child* by Peter Schrag and Diane Divoky (1975), are being drugged in the schools, and one-fourth of the American female population in the thirty-to-sixty age group uses psychoactive prescription drugs on a regular basis. Articles in popular magazines such as *Cosmopolitan* urge sufferers from depression to drop in to the local mental hospital for drugs or shock treatments, so that they can return to their jobs as quickly as possible. "The drug and the mental hospital", writes one political scientist, "have become the indispensable lubricating oil and reservicing factory needed to prevent the complete breakdown of the human engine".[7]

These figures are American in degree, but not in kind. Poland and Russia are world leaders in the consumption of hard liquor; the suicide rate in France has been growing steadily; in West Germany, the suicide rate doubled between 1966 and 1976.[8] The insanity of Los Angeles and Pittsburgh is archetypal, and the "misery index" has been climbing in Leningrad, Stockholm, Milan, Frankfurt and other cities since midcentury. If America is the frontier of the Great Collapse, the other industrial nations are not far behind.

It is an argument of this book that we are *not* witnessing a peculiar twist in the fortunes of postwar Europe and America, an aberration that can be tied to such late twentieth-century problems as inflation, loss of empire, and the like. Rather, we are witnessing the inevitable outcome of a logic that is already centuries old, and which is being played out in our own lifetime. I am not trying to argue that science is the cause of our predicament; causality is a type of historical explanation which I find singularly unconvincing. What I am arguing is that the scientific world view is *integral* to modernity, mass society, and the situation described above. It is *our* consciousness, in the Western industrial nations — uniquely so — and it is intimately bound up with the emergence of our way of life from the Renaissance to the present. Science, and our way of life, have been mutually reinforcing, and it is for this reason that the scientific world view has come under serious scrutiny at the same time that the industrial nations are beginning to show signs of severe strain, if not actual disintegration.

From this perspective, the transformations I shall be discussing, and the solutions I dimly perceive, are epochal, and this is all the more reason not to relegate them to the realm of theoretical abstraction. Indeed, I shall argue that such fundamental transformations impinge upon the details of our daily lives far more directly than the things we may think to be most urgent: this Presidential candidate, that piece of pressing legislation, and so on. There have been other periods in human history when the accelerated pace of transformation has had such an impact on individual lives, the Renaissance being the most recent example prior to the present. During such periods, the meaning of individual lives begins to surface as a disturbing problem, and people become preoccupied with the meaning of meaning itself. It appears a necessary concomitant of this preoccupation that such periods are characterized by a sharp increase in the incidence of madness, or more precisely, of what is seen to define madness.[9] For value systems hold us (*all* of us, not merely "intellectuals") together, and when these systems start to crumble, so do the individuals who live by them. The last sudden upsurge in depression and psychosis (or "melancholia", as these states of mind were then called) occurred in the sixteenth and seventeenth centuries, during which time it became increasingly difficult to maintain notions of salvation and God's interest in human affairs. The situation was ultimately stabilized by the emergence of the new mental framework of capitalism, and the new definition of reality based on the scientific mode of experiment, quantification, and technical mastery. The problem is that this whole constellation of factors — technological manipulation of the environment, capital accumulation based on it, notions of secular salvation that fueled it and were fueled by it — has apparently run its course. In particular, the modern scientific paradigm has become as difficult to maintain in the late twentieth century as was the religious paradigm in the seventeenth. The collapse of capitalism, the general dysfunction of institutions, the revulsion against ecological spoilation, the increasing inability of the scientific world view to explain the things that really matter, the loss of interest in work, and the statistical rise in depression, anxiety, and outright psychosis are all of a piece. As in the seventeenth century, we are again destabilized, cast adrift, floating. We have, as Dante wrote in the *Divine Comedy*, awoken to find ourselves in a dark wood.

What will serve to stabilize things today is fairly obscure; but it is a major premise of this book that because disenchantment is intrinsic to the scientific world view, the modern epoch contained, from its inception, an inherent instability that severely limited its ability to sustain itself for more than a few centuries. For more than 99 percent of human history, the world was enchanted and man saw himself as an integral part of it. The complete reversal of this perception in a mere four hundred years or so has destroyed the continuity of the human experience and the integrity of the human psyche. It has very nearly wrecked the planet as well. The only hope, or so it seems to me, lies in a reenchantment of the world.

Here, then, is the crux of the modern dilemma. We cannot go

back to alchemy or animism — at least that does not seem likely; but the alternative is the grim, scientistic, totally controlled world of nuclear reactors, microprocessors, and genetic engineering — a world that is virtually upon us already. *Some* type of holistic, or participating, consciousness and a corresponding sociopolitical formation have to emerge if we are to survive as a species. At this point, as I have said, it is not at all evident what this change will involve; but the implication is that a way of life is slowly coming into being which will be vastly different from the epoch that has so deeply colored, in fact created, the details of our lives. Robert Heilbroner has suggested that a time might come, perhaps two hundred years hence, when people will visit the Houston computer center or Wall Street as curious relics of a vanished civilization, but this will necessarily involve a dramatically altered perception of reality.

1 Russell Jacoby, *Social Amnesia* (Boston: Beacon Press, 1975), p. 63.
2 Herbert Marcuse, *One-Dimensional Man* (Boston: Beacon Press, 1964), pp. 9, 154.
3 Studs Terkel, *Working* (New York: Avon Books,1972).
4 Richard Sennett and Jonathan Cobb, *The Hidden Injuries of Class* (New York: Vintage Books, 1973), pp. 168ff.
5 The elaboration of this process is perhaps the greatest contribution of the Frankfurt School for Social Research, whose most familiar representative in the United States was Herbert Marcuse. A summary of their work may be found in Martin Jay, *The Dialectical Imagination* (Boston: Little, Brown, 1973). On the popular level, Vance Packard has provided much evidence for this view of the totally manipulated life in books such as *The Status Seekers, The Hidden Persuaders*, and several others.
6 Joseph A. Camilleri, *Civilization in Crisis* (Cambridge: Cambridge University Press, 1976), pp. 31-32. The incredible emphasis on sexual technique, as opposed to emotional content, is reflected in the voluminous proliferation of sex manuals in the last fifteen years, by now a multimillion dollar business.
7 Camilleri, *Civilization in Crisis*, p. 42. Information such as this can be collected, at this point, by merely reading daily newspapers and popular journals. My own sources include *Newsweek*, 8 January 1973 and 12 November 1979; *National Observer*, 6 March 1976; *San Francisco Examiner*, 24 March 1977 and 10 July 1980; *San Francisco Chronicle*, 29 march 1976 and 10 September 1979; *New York Times*, 16 March 1976; *Cosmopolitan*, September 1974; and a general survey of such articles provided in John and Paula Zerzan, "Breakdown", which was published in abridged form in the January 1976 issue of *Fifth Estate*. The quotation from Darold Treffert is from this pamphlet. For an extended critique of American drug use, see Richard Hughes and Robert Brewin, *The Tranquilizing of America* (New York: Harcourt Brace Jovanovich, 1979).
8 According to a 1972 Finnish study, Poland, the Soviet Union and Hungary are respectively first, second and third in world per capita consumption of hard liquor. See *San Francisco Chronicle*, 8 September 1978. My information on

French and German suicides comes from a 1979 report of San Francisco's Pacific News Service by Eve Pell, "Teenage Suicides Sweep Advanced Nations of the West."

9 Dr. Edward F. Foulks, a medical anthropologist at the University of Pennsylvania, has argued that madness may be a way by which the human species protects itself in such times of crisis, and hence that psychosis may be a form of cultural avant-garde (see the report on his work in the *New York Times*, 9 December 1975, p. 22, and the *National Observer*, 6 march 1976, p. 1). Much of the work of R. D. Laing points in this direction, and it has been a theme in a number of Doris Lessing's novels. See also Andrew Weil, *The National Mind* (Boston: Houghton Mifflin, 1972).

10 Robert Heilbroner, *Business Civilization in Decline* (New York: Norton, 1976), pp. 120-124.

4. What is the future of human culture with respect to technology? Is there a solution to the reality of being diminished by high tech?

Life is becoming more and more technicized, mediated by machines, modeled on mechanical processes. Heidegger said in the 1960s that philosophy has come to an end in the present epoch, replaced by cybernetics.

Yet the awareness of this unbearable condition grows apace with the barrenness that technology creates around us.

Realistic efforts at a solution will be drastic, far-reaching ones as reflected by a pessimism among most critics of high tech. The problem is profound, and pessimism may be a necessary part of the first step in sizing up our oppression.

There seem to be indications that a whole critique of modern life is involved here. As things get worse, more empty, that encompassing critique will make more and more sense.

Almost fifty years ago, Georges Bataille wrote of the nature and result of modern science in a way that also applies to technology:

"Science is a function that developed only after occupying the place of the destiny it was to have *served* . . . It is a paradox that a function could only be fulfilled on condition that it become an end in itself. The totality of sciences that man has at his disposal is due to this sort of fraud. But if it is true that the human domain has increased because of it, it has been at the cost of a crippled existence."

Perhaps we are now, finally, able to see this crippling more clearly, and can comprehend that a subjugation of outer nature, now so evident — and ghastly — was truly purchased at the cost of suppression of inner nature. The instrumental or utilitarian character of science and technology is a false notion; domination itself is found there.

If this indictment is vast, so are the measures we must take to remove its application from a world we would like to save and savour. Since the word is getting out — as evidenced by the four samples below — the real work may be our will to renewal and our desire for wholeness.

T. Fulano's poetic contribution verges on despair in its metaphor of our technicized culture as a 747 crashing to earth. Indian activist Russell Means' manifesto attracted much attention as an angry denunciation of "European" culture as a whole. Sally Gearhart goes so far as to recommend the non-reproduction of the human species as a last alternative to high tech's destructiveness, while Morris Berman suggests some of the mammoth changes in the social order that would be necessary for human culture to survive.

Civilization is like a jetliner

T. FULANO

Civilization is like a jetliner, noisy, burning up enormous amounts of fuel. Every imaginable and unimaginable crime and pollution had to be committed in order to make it go. Whole species were rendered extinct, whole populations dispersed. Its shadow on the waters resembles an oil slick. Birds are sucked into its jets and vaporized. Every part, as Gus Grissom once nervously remarked about space capsules before he was burned up in one, has been made by the lowest bidder.

Civilization is like a 747, the filtered air, the muzak oozing over the earphones, a phony sense of security, the chemical food, the plastic trays, all the passengers sitting passively in the orderly row of padded seats staring at Death on the movie screen. Civilization is like a jetliner, an idiot savant in the cockpit, manipulating computerized controls built by sullen wage workers, and dependent for his* directions on sleepy technicians high on amphetamines with their minds wandering to sports and sex.

Civilization is like a 747, filled beyond capacity with coerced volunteers — some in love with the velocity , most wavering at the abyss of terror and nausea, yet still seduced by advertising and propaganda. It is like a DC-10, so incredibly enclosed that you want to break through the tin can walls and escape, make your own way through the clouds, and leave this rattling, screaming fiend approaching its breaking point. The smallest error or technical failure leads to catastrophe, scattering your sad entrails like belated omens all over the runway, knocks you out of your shoes, breaks all your bones like egg shells.

(Of course civilization is like many other things besides jets — always *things* — a chemical drainage ditch, a woodland knocked down to lengthen an airstrip or to build a slick new shopping mall where people can buy salad bowls made out of exotic tropical trees which will be extinct next week, or perhaps a graveyard for cars, or a

*See "Questioning Technology. Questioning Patriarchy", p. 1.

suspension bridge which collapses because a single metal pin has shaken loose. Civilization is a hydra. There is a multitude of styles, colors, and sizes of Death to choose from.)

Civilization is like a Boeing jumbo jet because it transports people who have never experienced their humanity where they were, to places where they shouldn't go. In fact it mainly transports {businessmen} in suits with briefcases filled with charts, contracts, more mischief —{businessmen} who are identical everywhere and hence have no reason at all to be ferried about. And it goes faster and faster, turning more and more places into airports, the (un)natural habitat of {businessmen}

It is an utter mystery how it gets off the ground. It rolls down the runway, the blinking lights along the ground like electronic scar tissue on the flesh of the earth, picks up speed and somehow grunts, raping the air, working its way up along the shimmering waves of heat and the trash blowing about like refugees fleeing the bombing of a city. Yes, it is exciting, a mystery, when life has been evacuated and the very stones have been murdered.

But civilization, like the jetliner, this freak phoenix incapable of rising from its ashes, also collapses across the earth like a million bursting wasps, flames spreading across the runway in tentacles of gasoline, samsonite, and charred flesh. And always the absurd rubbish, Death's confetti, the fragments left to mock us lying along the weary trajectory of the dying bird — the doll's head, the shoes, eyeglasses, a beltbuckle.

Jetliners fall, civilizations fall, this civilization will fall. The gauges will be read wrong on some snowy day (perhaps they will fail). The wings, supposedly defrosted, will be too frozen to beat against the wind and the bird will sink like a millstone, first gratuitously skimming a bridge (because civilization is also like a bridge, from Paradise to Nowhere), a bridge laden, say, with commuters on their way to or from work, which is to say, to or from an airport, packed in their cars (wingless jetliners) like additional votive offerings to a ravenous Medusa.

Then it will dive into the icy waters of a river, the Potomac perhaps, or the River Jordon, or Lethe. And we will be inside, each one of us at our specially assigned porthole, going down for the last time, like dolls' heads encased in plexiglass.

Fighting Words on the Future of the Earth

RUSSELL MEANS

The only possible opening for a statement of this kind is that I detest writing. The process itself epitomizes the European concept of "legitimate" thinking; what is written has an importance that is denied the spoken. My culture, the Lakota culture, has an oral tradition, so I ordinarily reject writing. It is one of the white world's ways of destroying the cultures of non-European peoples, the imposing of an abstraction over the spoken relationship of a people.

So what you read here is not what I've written. It's what I've said and someone else has written down. I will allow this because it seems that the only way to communicate with the white world is through the dead, dry leaves of a book. I don't really care whether my words reach whites or not. They have already demonstrated through their history that they cannot hear, cannot see; they can only read (of course, there are exceptions, but the exceptions only prove the rule). I'm more concerned with American Indian people, students and others, who have begun to be absorbed into the white world through universities and other institutions. But even then it's a marginal sort of concern. It's very possible to grow into a red face with a white mind; and if that's a person's individual choice, so be it, but I have no use for them. This is part of the process of cultural genocide being waged by Europeans against American Indian peoples today. My concern is with those American Indians who choose to resist this genocide, but who may be confused as to how to proceed.

(You notice I use the term *American Indian* rather than *Native American* or *Native indigenous people* or *Amerindian* when referring to my people. There has been some controversy about such terms, and frankly, at this point, I find it absurd. Primarily it seems that *American Indian* is being rejected as European in origin — which is true. But *all* the above terms are European in origin; the only non-European way is to speak of Lakota — or, more precisely, of Oglala, Brulé, etc. — and of the Diné, the Miccosukee and all the rest of the several hundred correct tribal names.

(There is also some confusion about the word *Indian*, a mistaken belief that it refers somehow to the country, India. When Columbus washed up on the beach in the Caribbean, he was not looking for a country called India. Europeans were calling that country Hindustan in 1492. Look it up on the old maps. Columbus called the tribal people he met "Indio", from the Italian *in dio*, meaning "in God".)

It takes a strong effort on the part of each American Indian *not* to become Europeanized. The strength for this effort can only come from the traditional ways, the traditional values that our elders retain. It must come from the hoop, the four directions, the relations; it cannot come from the pages of a book or a thousand books. No European can ever teach a Lakota to be Lakota, a Hopi to be Hopi. A master's degree in "Indian Studies" or in "education" or in anything else cannot make a person into a human being or provide knowledge into the traditional ways. It can only make you into a mental European, an outsider.

I should be clear about something here, because there seems to be some confusion about it. When I speak of Europeans or mental Europeans, I'm not allowing for false distinctions. I'm not saying that on the one hand there are the by-products of a few thousand years of genocidal, reactionary, European intellectual development which is bad; and on the other hand there is some new revolutionary intellectual development which is good. I'm referring here to the so-called theories of Marxism and anarchism and "leftism" in general. I don't believe these theories can be separated from the rest of the European intellectual tradition. It's really just the same old song.

The process began much earlier. Newton, for example, "revolutionized" physics and the so-called natural sciences by reducing the physical universe to a linear mathematical equation. Descartes did the same thing with culture. John Locke did it with politics, and Adam Smith did it with economics. Each one of these "thinkers" took a piece of the spirituality of human existence and converted it into a code, an abstraction. They picked up where Christianity ended; they "secularized" Christian religion, as the "scholars" like to say — and in doing so they made Europe more able and ready to act as an expansionist culture. Each of these intellectual revolutions served to abstract the European mentality even further, to remove the wonderful complexity and spirituality from the universe and replace it with a logical sequence: one, two, three, Answer!

This is what has come to be termed "efficiency" in the European mind. Whatever is mechanical is perfect; whatever seems to work at the moment — that is, proves the mechanical model to be the right one — is considered correct, even when it is clearly untrue. This is why "truth" changes so fast in the European mind; the answers which result from such a process are only stop-gaps, only temporary, and must be continuously discarded in favor of new stop-gaps which support the mechanical models and keep them (the models) alive.

Hegel and Marx were heirs to the thinking of Newton, Descartes, Locke and Smith. Hegel finished the process of secularizing theology — and that is put in his own terms — he secularized the religious thinking through which Europe understood the universe. Then Marx put Hegel's philosophy in terms of "materialism", which is to say that Marx despiritualized Hegel's work altogether. Again, this is in Marx's own terms. And this is now seen as the future revolutionary potential of Europe. Europeans may see this as revolutionary, but American Indians see it simply as still more of that same old European conflict between *being* and *gaining*. The intellectual roots of a new Marxist form of European imperialism lie in Marx's — and his followers' — links to the tradition of Newton, Hegel and the others.

Being is a spiritual proposition. *Gaining* is a material act. Traditionally, American Indians have always attempted to *be* the best people they could. Part of that spiritual process was and is to give away wealth, to discard wealth in order *not* to gain. Material gain is an indicator of false status among traditional people, while it is "proof that the system works" to Europeans. Clearly, there are two completely opposing views at issue here, and Marxism is very far over to the other side from the American Indian view. But let's look at a major implication of this; it is not merely an intellectual debate.

The European materialist tradition of despiritualizing the universe is very similar to the mental process which goes into dehumanizing another person. And who seems most expert at dehumanizing other people? And why? Soldiers who have seen a lot of combat learn to do this to the enemy before going back into combat. Murderers do it before going out to commit murder. Nazi

SS guards did it to concentration camp inmates. Cops do it. Corporation leaders do it to the workers they send into uranium mines and steel mills. Politicians do it to everyone in sight. And what the process has in common for each group doing the dehumanizing is that it makes it all right to kill and otherwise destroy other people. One of the Christian commandments says, "Thou shalt not kill", at least not humans, so the trick is to mentally convert the victims into nonhumans. Then you can proclaim violation of your own commandment as a virtue.

In terms of the despiritualization of the universe, the mental process works so that it becomes virtuous to destroy the planet. Terms like *progress* and *development* are used as cover words here, the way *victory* and *freedom* are used to justify butchery in the dehumanization process. For example, a real-estate speculator may refer to "developing" a parcel of ground by opening a gravel quarry; *development* here means total, permanent destruction, with the earth itself removed. But European logic has *gained* a few tons of gravel with which more land can be "developed" through the construction of road beds. Ultimately, the whole universe is open — in the European view — to this sort of insanity.

Most important here, perhaps, is the fact that Europeans feel no sense of loss in all this. After all, their philosophers have despiritualized reality, so there is no satisfaction (for them) to be gained in simply observing the wonder of a mountain or a lake or a people *in being*. No, satisfaction is measured in terms of gaining material. So the mountain becomes gravel, and the lake becomes coolant for a factory, and the people are rounded up for processing through the indoctrination mills Europeans like to call schools.

But each new piece of that "progress" ups the ante out in the real world. Take fuel for the industrial machine as an example. Little more than two centuries ago, nearly everyone used wood — a replenishable, natural item — as fuel for the very human needs of cooking and staying warm. Along came the Industrial Revolution and coal became the dominant fuel, as production became the social imperative for Europe. Pollution began to become a problem in the cities, and the earth was ripped open to provide coal whereas wood had always simply been gathered or harvested at no great expense to the environment. Later, oil became the major fuel, as the technology of production was perfected through a series of scientific "revolutions". Pollution increased dramatically, and nobody yet knows what the environmental costs of pumping all that oil out of the ground will really be in the long run. Now there's an "energy crisis", and uranium is becoming the dominant fuel.

Capitalists, at least, can be relied upon to develop uranium as fuel only at the rate at which they can show a good profit. That's their ethic, and maybe that will buy some time. Marxists, on the other hand, can be relied upon to develop uranium fuel as rapidly as possible simply because it's the most "efficient" production fuel available. That's *their* ethic, and I fail to see where it's preferable. Like I said, Marxism is right smack in the middle of the European tradition. It's the same old song.

There's a rule of thumb which can be applied here. You cannot judge the real nature of a European revolutionary doctrine on the basis of the changes it proposes to make within the European power structure and society. You can only judge it by the effects it will have on non-European peoples. This is because every revolution in European history has served to reinforce Europe's tendencies and abilities to export destruction to other peoples, other cultures and the environment itself. I defy anyone to point out an example where this is not true.

So now we, as American Indian people, are asked to believe that a "new" European revolutionary doctrine such as Marxism will reverse the negative effects of European history on us. European power relations are to be adjusted once again, and that's supposed to make things better for all of us. But what does this really mean?

Right now, today, we who live on the Pine Ridge Reservation are living in what white society has designated a "National Sacrifice Area". What this means is that we have a lot of uranium deposits here, and white culture (not us) needs this uranium as energy production material. The cheapest, most efficient way for industry to extract and deal with the processing of this uranium is to dump the waste by-products right here at the digging sites. Right here where we live. This waste is radioactive and will make the entire region uninhabitable forever. This is considered by industry, and by the white society that created this industry, to be an "acceptable" price to pay for energy resource development. Along the way they also plan to drain the water table under this part of South Dakota as part of the industrial process, so the region becomes doubly uninhabitable. The same sort of thing is happening down in the land of the Navajo and Hopi, up in the land of the Northern Cheyenne and Crow, and elsewhere. Thirty percent of the coal in the West and half of the uranium deposits in the US have been found to lie under reservation land, so there is no way this can be called a minor issue.

We are resisting being turned into a National Sacrifice Area. We are resisting being turned into a national sacrifice people. The costs of this industrial process are not acceptable to us. It is genocide to dig uranium here and drain the water table — no more, no less.

Now let's suppose that in our resistance to extermination we begin to seek allies (we have). Let's suppose further that we were to take revolutionary Marxism at its word: that it intends nothing less than the complete overthrow of the European capitalist order which has presented this threat to our very existence. This would seem to be a natural alliance for American Indian people to enter into. After all, as the Marxists say, it is the capitalists who set us up to be a national sacrifice. This is true as far as it goes.

But, as I've tried to point out, this "truth" is very deceptive. Revolutionary Marxism is committed to even further perpetuation and perfection of the very industrial process which is destroying us all. It offers only to "redistribute" the results — the money, maybe — of this industrialization to a wider section of the population. It offers to take wealth from the capitalists and pass it around; but in order to do so, Marxism must maintain the industrial system. Once again, the power relations within European society will have to be altered, but once again the effects upon American Indian peoples here and non-Europeans elsewhere will remain the same. This is much the same as when power was redistributed from the church to private business during the so-called bourgois revolution. European society changed a bit, at least superficially, but its conduct toward non-Europeans continued as before. You can see what the American Revolution of 1776 did for American Indians. It's the same old song.

Revolutionary Marxism, like industrial society in other forms, seeks to "rationalize" all people in relation to industry — maximum industry, maximum production. It is a materialist doctrine that despises the American Indian spiritual tradition, our cultures, our lifeways. Marx himself called us "precapitalists" and "primitive". *Precapitalist* simply means that, in his view, we would eventually discover capitalism and become capitalists; we have always been economically retarded in Marxist terms. The only manner in which American Indian people could participate in a Marxist revolution would be to join the industrial system, to become factory workers, or "proletarians" as Marx called them. The man was very clear about the fact that his revolution could occur only through the struggle of the proletariat, that the existence of a massive industrial system is a precondition of a successful Marxist society.

I think there's a problem with language here. Christians, capitalists, Marxists. All of them have been revolutionary in their own minds, but none of them really mean revolution. What they really mean is a continuation. They do what they do in order that European culture can continue to exist and develop according to its needs.

So, in order for us to *really* join forces with Marxism, we American Indians would have to accept the national sacrifice of our homeland; we would have to commit cultural suicide and become industrialized and Europeanized.

At this point, I've got to stop and ask myself whether I'm being too harsh. Marxism has something of a history. Does this history bear out my observations? I look to the process of industrialization in the Soviet Union since 1920 and I see these Marxists have done what it took the English Industrial Revolution 300 years to do; and the Marxists did it in 60 years. I see that the territory of the USSR used to contain a number of tribal peoples and that they have been crushed to make way for the factories. The Soviets refer to this as "The National Question", the question of whether the tribal peoples had the right to exist as peoples; and they decided the tribal peoples were an acceptable sacrifice to industrial needs. I look to China and I see the same thing. I look to Vietnam and I see Marxists imposing an industrial order and rooting out the indigenous tribal mountain people.

I hear a leading Soviet scientist say that when uranium is exhausted alternatives will be found. I see Vietnamese taking over a nuclear plant abandoned by the US military. Have they dismantled and destroyed it? No, they are using it. I see China exploding nuclear bombs, developing uranium reactors and preparing a space program in order to colonize and exploit the planets the same as the Europeans colonized and exploited this hemisphere. It's the same old song, but maybe with a faster tempo this time.

The statement of the Soviet scientist is very interesting. Does he know what this alternative energy source will be? No, he simply has faith. Science will find a way. I hear revolutionary Marxists saying that the destruction of the environment, pollution and radiation will all be controlled. And I see them act upon their words. Do they know *how* these things will be controlled? No. they simply have faith.

Science will find a way. Industrialization is fine and necessary. How do they know this? Faith. Science will find a way. Faith of this sort has always been known in Europe as religion. Science has become the new European religion for both capitalists and Marxists; they are truly inseparable; they are part and parcel of the same culture. So, in both theory and practice, Marxism demands that non-European peoples give up their values, their traditions, their cultural existence altogether. We will all be industrialized science addicts in a Marxist society.

I do not believe that capitalism itself is really responsible for the situation in which American Indians have been declared a national sacrifice. No, it is the European tradition; European culture itself is responsible. Marxism is just the latest continuation of this tradition, not a solution to it. To ally with the very same forces that declare us an acceptable cost.

There is another way. There is the traditional Lakota way and the ways of the other American Indian peoples. It is the way that knows that humans do not have the right to degrade Mother Earth, that there are forces beyond anything that the European mind has conceived, that humans must be in harmony with *all* relations or the relations will eventually eliminate the disharmony. A lopsided emphasis on humans by humans — the Europeans' arrogance of acting as though they were beyond the nature of all related things — can only result in a total disharmony and a readjustment which cuts arrogant humans down to size, gives them a taste of that reality beyond their grasp on control and restores the harmony. There is no need for a revolutionary theory to bring this about; it's beyond human control. The nature peoples of this planet know this so they do not theorize about it. Theory is an abstract; our knowledge is real.

Distilled to its basic terms, European faith — including the new faith in science — equals a belief that man is God. Europe has always sought a Messiah, whether that be the man Jesus Christ or the man Karl Marx or the man Albert Einstein. American Indians know this to be totally absurd. Humans are the weakest of all creatures, so weak that other creatures are willing to give up their flesh that we may live. Humans are able to survive only through the exercise of rationality since they lack the abilities of other creatures to gain food through the use of fang and claw.

But rationality is a curse since it can cause humans to forget the natural order of things in ways other creatures do not. A wolf never forgets his or her place in the natural order. American Indians can.

Europeans almost always do. We pray our thanks to the deer, our relations, for allowing us their flesh to eat; Eurpeans simply take the flesh for granted and consider the deer inferior. After all, Europeans consider themselves godlike in their rationalism and science. God is the Supreme Being; all else *must* be inferior.

All European tradition, Marxism included, has conspired to defy the natural order of all things. Mother Earth has been abused, the powers have been abused, and this cannot go on forever. No theory can alter that simple fact, Mother Earth will retaliate, the whole environment will retaliate, and the abusers will be eliminated. Things come full circle, back to where they started. *That's* revolution. And that's a prophecy of my people, of the Hopi people and of other correct peoples.

American Indians have been trying to explain this to Europeans for centuries. But, as I said earlier, Europeans have proven themselves unable to hear. The natural order will win out, and the offenders will die out, the way deer die when they offend the harmony by overpopulating a given region. It's only a matter of time until what Europeans call "a major catastrophe of global proportions" will occur. It is the role of American Indian peoples, the role of all natural beings, to survive. A part of our survival is to resist. We resist not to overthrow a government or to take political power, but because it is natural to resist extermination, to survive. We don't want power over white institutions; we want white institutions to disappear. *That's* revolution.

American Indians are still in touch with these realities — the prophecies, the traditions of our ancestors. We learn from the elders, from nature, from the powers. And when the catastrophe is over, we American Indian peoples will all be here to inhabit the hemisphere. I don't care if it's only a handful living in the Andes. American Indian people will survive; harmony will be established. *That's* revolution.

At this point, perhaps, I should be very clear about another matter, one which should already be clear as a result of what I've said. But confusion breeds easily these days, so I want to hammer home this point. When I use the term *European*, I'm not referring to a skin color or a particular genetic structure. What I'm referring to is mind-set, a world view that is a product of the development of European culture. People are not genetically encoded to hold this outlook; they are *acculturated* to it. The same is true for American Indians or for the members of any culture.

It is possible for an American Indian to share European values, a European world view. We have a term for these people; we call them "apples" — red on the outside (genetics) and white on the inside (their values). Other groups have similar terms: Blacks have their "oreos": Hispanics have "coconuts" and so on. And, as I said before, there *are* exceptions to the white norm: people who are white on the outside, but not white inside. I'm not sure what term should be applied to them other than "human beings".

What I'm putting out here is not a racial proposition but a cultural proposition. Those who ultimately advocate and defend the realities of European culture and its industrialisation are my enemies. Those who resist it, who struggle against it, are my allies, the allies of American Indian people. And I don't give a damn what their skin colour happens to be. *Caucasian* is the white term for the white race; *European* is an outlook I oppose.

The Vietnamese Communists are not exactly what you might consider genetic Caucasians, but they are now functioning as mental Europeans. The same holds true for Chinese Communists, for Japanese capitalists or Bantu Catholics or Peter "MacDollar" down at the Navajo Reservation or Dickie Wilson up here at Pine Ridge. There is no racism involved in this, just an acknowledgement of the mind and spirit that make up culture.

In Marxist terms I suppose I'm a "cultural nationalist". I work first with my people, the traditional Lakota people, because we hold a common world view and share an immediate struggle. Beyond this, I work with other traditional American Indian peoples, again because of a certain commonality in world view and form of struggle. Beyond that, I work with anyone who has experienced the colonial oppression of Europe and who resists its cultural and industrial totality. Obviously, this includes genetic Caucasians who struggle to resist the dominant norms of European culture. The Irish and the Basques come immediately to mind, but there are many others.

I work primarily with my own people, with my own community. Other people who hold non-European perspectives should do the same. I believe in the slogan, "Trust your brother's vision", although I'd like to add sisters into the bargain. I trust the community and the culturally based vision of all the races that naturally resist industrialization and human extinction. Clearly, individual whites can share in this, given only that they have reached the awareness that continuation of the industrial imperatives of Europe is not a vision, but species suicide. White is one of the sacred colors of the Lakota people — red, yellow, white and black. The four directions. The four seasons. The four periods of life and aging. The four races of humanity. Mix red, yellow, white and black together and you get brown, the colour of the fifth race. This is a natural ordering of things. It therefore seems natural to me to work with all races, each with its own special meaning, identity and message.

But there is a peculiar behaviour among most Caucasians. As soon as I became critical of Europe and its impact on other cultures, they become defensive. They begin to defend themselves. But I'm not attacking them personally; I'm attacking Europe. In personalizing my observations on Europe they are personalizing European culture, identifying themselves with it. By defending themselves in *this* context, they are ultimately defending the death culture. This is a confusion which must be overcome, and it must be overcome in a hurry. None of us have energy to waste in such false struggles.

Caucasians have a more positive vision to offer humanity than European culture. I believe this. But in order to attain this vision it is necessary for Caucasians to step outside European culture — alongside the rest of humanity — to see Europe for what it is and what it does.

To cling to capitalism and Marxism and all the other "isms" is simply to remain within European culture. There is no avoiding this basic fact. As a fact, this constitutes a choice. Understand that the choice is based on culture, not race. Understand that to choose European culture and industrialism is to choose to be my enemy. And understand that the choice is yours, not mine.

This leads me back to address those American Indians who are drifting through the universities, the city slums and other European institutions. If you are there to learn to resist the oppressor in accordance with your traditional ways, so be it. I don't know how you manage to combine the two, but perhaps you will succeed. But retain your sense of reality. Beware of coming to believe the white world now offers solutions to the problems it confronts us with. Beware, too, of allowing the words of native people to be twisted to the advantage of our enemies. Europe invented the practice of turning words around on themselves. You need only look to the treaties between American Indian peoples and various European governments to know that this is true. Draw your strength from who you are.

A culture which regularly confuses revolution with continuation, which confuses science and religion, which confuses revolt with resistance, has nothing helpful to teach you and nothing to offer you as a way of life. Europeans have long since lost all touch with reality, if ever they were in touch with it. Feel sorry for them if you need to, but be comfortable with who you are as American Indians.

So, I suppose to conclude this, I should state clearly that leading anyone toward Marxism is the last thing on my mind. Marxism is as alien to my culture as capitalism and Christianity are. In fact, I can say I don't think I'm trying to lead anyone toward anything. To some extent I tried to be a "leader", in the sense that the white media like to use that term, when the American Indian Movement was a young organization. This was a result of a confusion I no longer have. You cannot be everything to everyone. I do not propose to be used in such a fashion by my enemies. I am not a leader. I *am* an Oglala Lakota patriot. That is all I want and all I need to be. And I am very comfortable with who I am.

An End to Technology

SALLY GEARHART

Conditioned as we have become to epistemology and the values of Western science and technology, we grow more and more *dependent* upon these enterprises. With its tendency to speed and size, technology increases the dangers of that dependency by leaving greater and greater quantities of people helpless when technology fails. The nuclear meltdown at Three Mile Island and the blackouts / brownouts of large metropolitan areas stand as prime examples of such helplessness. Further, computer or airline breakdowns frequently leave hosts of irate consumers stranded; the Nestlé Corporation consciously weans African babies from traditional nutrition only to see them die when their own commercial formulas prove fatal (*AFSC Women's Newsletter* 1982). As technology proliferates and population increases, we may discover that we have cut off even our escapes from the growing dependency.

Not only do we cultivate a dangerous dependency on technology; as well, Western science would have us all stand in an adversarial relationship to our environment. If we are to study and control nature, then we must distinguish it from ourselves, *alienate* ourselves from it. The limited epistemology and the questionable values of science allow us to make that separation. As a consequence, we can "do to" other entities what we would never think of doing to ourselves. If we perceive an "other" to be a part of ourselves, then we deprive ourselves of the luxuries we have become so dependent upon.

One question follows on this condition: if technology is not neutral, if each time we pick up a tool we alienate ourselves by the capacity of the tool from the material or the task we are addressing ourselves to, then can we justify the use of any tool? I presently contend that we cannot, not if a human being is the user of the tool. The culprit is our human consciousness which has shown a tendency to misuse the information supplied to it by our intelligence. It is this consciousness which has acceded to the Faustian temptations of "knowledge" and has made us into the technological animal, the

83

animal that consciously and systematically manipulates its environment. Thus, by picking up a tool, any tool, we commit ourselves to the precepts of Western science or some painfully like them; complete with the dependency and alienation that accrue to them. While I would like to hope that we can undo Western science, I seriously doubt that that can happen. The train rushes downhill too fast.

The gift of intelligence deserves a more responsible and sensitive deployment than human beings have given it. We might have asked of the use of tools, "Whom does it help, whom does it hurt?" We might have had the bigger picture, the longer view. We might have observed that personal, class or ethnic loyalties are far too limited; so are national loyalties or even loyalties to this or that multinational corporation. We might have observed, out of the gift of intelligence, that the deepest loyalty of any species must ultimately be to its environment, its home, the original source of its life, and in the cycle of matter — the place to which it will eventually return.

To be sure, some groups of human beings have related well to the earth, have developed a consciousness worthy of our intellectual gifts. But, precisely because they refused technology, they became vulnerable to those who didn't. Technology leads to power, domination, control. Our failed consciousness in our relationship to the planet will inevitably, alas, prevail. The one rotten apple spoils the barrel. Until the worst of us is reformed, the whole species seems doomed; our nature seems to be to destroy whole portions of our self.

We have used our consciousness not to harmonize with nature but to change and conquer it. The earth seems now to be giving unmistakable signals that it can no longer bear the weight, either of our numbers or of our arrogance. Nor, apparently, can we change our consciousness or stop the pell-mell rush to planetary destruction. Only one alternative remains.

An extraterrestrial observing our polluted and diseased planet would have to conclude that homo sapiens, the inventor of technology, was an evolutionary blunder and should now silently fold its wings and steal away. I agree with that cosmic observer. From the point of view of our fellow species and the earth itself, the best that can happen is that human beings never conceive another child, that the child being conceived at this very moment be the last human being ever to exist.

Granted, the intervention of some outside force whose magic wand sterilizes the entire race does not gather many bets. Even slimmer is the likelihood that human beings will develop the consciousness necessary for a voluntary cessation of reproduction. Those odds, however, in no way reduce the integrity or the good sense of the proposal. The idea suggests absolutely no bloodshed, no destruction of property or reputation. There will simply be no more of us. Ever. To those who cry "That flies in the face of our drive to preserve the species", I answer that recent history does not document even the slightest inclination on our part toward species preservation, but, rather, records our determination to poison and asphyxiate ourselves. The greater virtue of my solution rests in the fact that when we go we will have the decency not to take the rest of the planet with us.

What would we lose by ceasing to be as a species? What would we lose personally? One possibility, of course, if we knew we were to have no more of us, takes the form of a trigger-happy and hysterical species that pushes all the nuclear buttons right away just to be sure we don't die alone. More optimistically, we might change character pretty quickly if we knew we were the last of our kind. We might stop killing each other; human life might take on some of the dignity that all life should have; we might appreciate our children more and begin to see what we could have been, how with a different set of values and epistemological assumptions we might have managed it all differently. The last days of homo sapiens might, after all, reveal some hidden life-loving capacity, quickened only by the hope that the earth can now restore itself from its long and difficult relationship with the human race.

This alternative affirms a far-sighted and humble participation in environmental affairs. Although it condemns the species, it need in no way diminish our regard for some remarkable individuals — perhaps including ourselves — who are a part of the species. I find in this solution of our species suicide an integrity of the calibre that humanists have long claimed is possible for the human race, though such a solution may not seem at first glance to be expressive of such integrity. If some still ask "Why?" I suggest that the burden of proof has shifted, that in terms of our biosphere the question is, "Why not?"

REFERENCES

American Friends Service Committee Women's Newsletter, 1982, 1 (1 and 2): 82.
Bogen, J., 1972. The other side of the brain. *Bulletin of the Los Angeles Neurological Society*, 4: 1-3.

The Re-enchantment of the World — II

MORRIS BERMAN

Human culture will come to be seen more as a category of natural history, "a semipermeable membrane between man and nature".[1] Such a society will be preoccupied with fitting into nature rather than attempting to master it. The goal will be "not to *rule* a domain, but to *release* it"; to have, once again, "clean air, clean clear-running rivers, the presence of Pelican and Osprey and Gray Whale in our lives; salmon and trout in our streams; unmuddied language and good dreams."[2] Technology will no longer pervade our consciousness and its presence will be more in the form of crafts and tools, things that lie *within* our control rather than the reverse.[3] We will no longer depend on the technological fix, whether in medicine, agriculture, or anything else, but instead favor solutions that are long-term and address themselves to causes rather than symptoms.

Politically, there will be a tremendous emphasis on decentralization, which will extend to all the institutions of society and be recognized as a prerequisite to planetary culture itself. Decentralization implies that institutions are small-scale and subject to local control, and that political structures are regional and autonomous. Characteristic of such decentralization are community hospitals and food cooperatives, the cultivation of neighbourhood spirit and autonomy, and the elimination of such destroyers of community such as television, automobiles, and expressways. Mass production will yield to craftsmanship, agribusiness to small, organic, labor-intensive farming, and centralized energy sources — especially nuclear power plants — to renewable energy options appropriate to their own regions. Mass education centers teaching essentially one type of knowledge as preparation for a career will be replaced by direct apprenticeship, in the form of a lifelong education that follows one's changing interests. One will not have a career, but a *life*. The blight of suburbs and urban sprawl, truly the antithesis of

86

*See "Questioning Technology, Questioning Patriarchy," p. 1.

city life, will be replaced by a genuine city culture, one native to its own region rather than reflecting an international world of mass communication. The city will once again become a center of life and pleasure, an *agora* (that fine Greek word), a market place and meeting place, Philippe Ariès' "medley of colors'. People will live closer to their work, and in general there will not be much distinction between work, life and leisure[4] . . .

How are we going to get there? From the present vantage point, the vision of a future in which fact and value are once again reunited, in which men and women have control over their own destinies, and in which ego-consciousness is more reasonably situated within a larger context of Mind, seems utopian in the extreme. Yet as Octavio Paz observed, the only alternative is suicide. Western industrial society has reached the limits of its own deutero-learning, and much of it now is in the midst of the social analogue of either madness or creativity, that is, re-creation (Learning III). Given this situation, how utopian is such a vision? Of course, if one believes that only violent revolution produces substantive change, and that such a transformation can be accomplished within a few decades, then planetary culture does not have much of a chance. If, however, we are talking of a change on the scale of the disintegration of the Roman Empire, such as has been suggested by Theodore Roszak, Willis Harman, and Robert Heilbroner, among others, then our utopian vision starts to appear increasingly realistic.[5] In fact, one of the most effective agents of this set of changes is the decay of advanced industrial society itself. Thus Percival Goodman writes in *The Double E* that the conserver society will not come about because of voluntary effort, but because the planet simply cannot support the world of an ever-expanding Gross National Product. Industrial economies are starting to contract. We may choose to make a virtue our of what has been called "Buddhist economics", but we shall have to return to a steady-state economy whether we like it or not.[6]

Social change is also being generated by millions of individuals who have no interest in change per se, but have effectively undertaken an "inner migration", or withdrawal. Both Harman and Heilbroner have pointed to the fact that the industrial economies are going to face a severe economic crunch at the very time that their workers, both blue and white collar, have found their work so devoid of intrinsic value that they are increasingly finding meaning elsewhere, and privately withdrawing their allegiance from their jobs. The Protestant work ethic, the spiritual support of our present way of life, will not be there when the economy needs it most. A 1975 report of the Trend Analysis Program of the American Institute of Life Insurance predicts a weakening of "industrial era philosophy" during the next two decades, with concomitant worker alienation, slowdowns, sabotage, and riots. "We may", concludes the report, "be somewhere in the middle of a turbulent transition to a new, or at least somewhat different, culture," beginning about 1990.[7]

On the political level, decay will probably take the form of the breakup of the nation-state in favor of small, regional units. This trend, sometimes called political separatism, devolution, or balkanization, is by now quite widespread in all industrial societies. The number of new nations has risen dramatically since 1945, and other societies are beginning to fragment into provincial and sectarian sub-units. Leopold Kohr predicted this trend (enthusiastically) as early as 1957 in his book, *The Breakdown of Nations*; official culture, such as *Harper's*, is now terrified of it. More soberly, a group of about 200 European experts, in the book *Europe 2000*, sees the revolt of a regional periphery as very likely.[8] There are now strong separatist movements not only in the United States (Northern California, Upper Michigan, Idaho's Panhandle), but in Scotland, Brittany, Pays Basque, and Corsica; and many other countries are also experiencing strong regional sentiments, so much so that the Europe of 2000 AD may well look like a mosaic of very small states. This process represents a reversion to original political boundaries that existed prior to the rise of modern nation-states: not France, but Burgundy, Picardy, Normandy, Alsace, and Lorraine; not Germany, but Bavaria, Baden, Hesse, Hanover; not Spain, but Valencia, Aragon, Catalonia, Castile; and so on. In general, writes Peter Hall, what at all levels

used to be called separatism and is now usually called regionalism — fundamentally the desire and willingness to assume more direct control over one's own destiny — is perhaps the strongest political drive now operating: it is the main cause of the "crisis of authority" and the weakening of centralized control.[9]

Holistic society is thus coming upon us from a variety of sources that cut across the traditional left-right political axis. Feminism, ecology, ethnicity, and transcendentalism (religious renewal), which ostensibly have nothing in common politically, may be converging toward a common goal. These holistic movements do not represent a single social class, nor can they even be analyzed in such terms, for by and large they represent the repressed "shadows" of industrial civilization: the feminine, the wilderness, the child, the body, the creative mind and heart, the occult, and the peoples of the non-urban, regional peripheries of Europe and North America — regions that have never bought into the ethos of the industrial heartland and never will. If there is any bond among the elements of this "counterculture", it is the notion of recovery. Their goal is the recovery of our bodies, our health, our sexuality, our natural environment, our archaic traditions, our unconscious mind, our rootedness in the land, our sense of community, and connectedness to one another. What they advocate is not merely a program of "no growth" or industrial slowdown, but the direct attempt to get back from the past what we lost during the last four centuries; to go backward in order to go forward. In a word, they represent the attempt to recover our future.

What is remarkable in many of these developments, also, is the attempt to create a politics that does not substitute one set of rulers for another, or even one political structure for another, but which reflects the basic needs of mind, body, sexuality, community, and the like. The goal, notes that ancient Chinese oracle, the *I Ching*, is

a satisfactory political or social organization of mankind. [Therefore] we must go down to the very foundations of life. For any merely superficial ordering of life that leaves its deepest needs unsatisfied is as ineffectual as if no attempt at order had ever been made.[10]

In various ways, this has become the goal of all holistic politics; a politics that would be the end of politics, at least as we know it today.

If all these changes, or even a third of them, came to pass, the anomie of the modern era would surely be a closed chapter in our history. Such a planetary culture would of neccessity erase our contemporary feeling of homelessness, and the sense that our personal reality is at odds with official reality. The infinite spaces whose silence terrified Pascal may appear to men and women of the future as extensions of a biosphere that is nurturing and benevolent. Meaning will no longer be something that must be found and imposed on an absurd universe; it will be given, and, as a result, men and women will have a feeling of cosmic connectedness, of belonging to a larger pattern. Surely, such a world represents salvation, but only in the sense that there is no need to be saved in the first place. A loss of interest in the traditional opiates would likely follow, and even psychoanalysis would be seen as superfluous. What would be worshipped, if anything, is ourselves, each other, and *this earth* — our *home*, the body of us all that makes our lives possible.

1 Jerry Gorsline and Linn House, "Future Primitive".
2 Murray Bookchin, *Post-Scarcity Anarchism*, p. 78; Gary Snyder, "Four Changes", p. 94.
3 There is by now a large literature on what is called "appropriate technology", or "technology with a human face". Two of the most well-known works are Ivan Illich, *Tools for Conviviality* (New York: Harper & Row. 1973), and Schumacher, *Small is Beautiful*.
4 Murray Bookchin, *The Limits of the City* (New York: Harper & Row, 1973); Lewis Mumford, *The Culture of Cities* (New York: Harcourt, Brace and Company, 1938). The quote from Ariès is in *Centuries of Childhood*, trans. Robert Baldick (New York: Vintage Books, 1962), p. 414.
5 Roszak's work, especially *Unfinished Animal* (New York: Harper & Row, 1975), *Where the Wasteland Ends* (Garden City, NY: Doubleday Achor, 1972), and *Person / Planet* (Garden City, NY: Doubleday, 1978), is premised on the Roman Empire model. Cf. Harman, *Incomplete Guide* and Robert L. Heilbroner, *Business Civilization in Decline*. (New York: Norton, 1976).
6 Percival Goodman, *The Double E* (Garden City, NY: Doubleday Anchor, 1977).
7 "A Future That Means Trouble", *San Francisco Chronicle*, 22 December 1975.
8 Leopold Kohr, *The Breakdown of Nations*, Kevin Phillips, "The Balkanization of America", *Harper's*, May 1978, pp. 37-47, Peter Hall, *Europe 2000*, esp. pp. 22-27 (in general all of the trends I have sketched in the vision of a planetary culture are laid out in this book, including some of the sources for change). See also Zwerin, *Case for the Balkanization of Practically Everyone*.
9 Hall, *Europe 2000*, p. 167.
10 *I Ching*, trans. Richard Wilhelm and Cary F. baynes, 3rd ed. (Princeton: Princeton University Press, 1967), p. 186 (Hexagram 48, The Well).

II

COMPUTERS
and the Informed Individual

---

5. Are computers a force for increased individual autonomy — or a route to a new totalitarianism?

Computer and *count* derive from the same Latin words (con + putare), to reckon. Counting is subconsciously identified with fingers and toes and manageable numbers. The computer, though, has been termed "an abacus on amphetamines". Its ability to perform arithmetical chores far outstrips human skill in speed and accuracy, just as the dishwasher gets dishes cleaner, cars are faster than legs, and typing is easier to read than handwriting.

Computers used to be sold as labor-saving devices; nowadays thay are more commonly touted as indispensible tools that give people a head start in the race of life. Kids need them to make it from kindergarten to college; executives need them so as not to waste precious seconds or make decisions based on insufficient data. Countries must have them to keep up with other countries, especially militarily.

An increasing percentage of office work is computer-based. Computerization has turned many offices into the information version of an assembly line, where a discrete unit is processed only to be replaced by the next discrete unit, with no let-up, no breather.

In this section, David Burnham focuses on the increased power modern bureaucracies have to override personal privacy via computers; Abbe Mowoshowitz sees a diminution of consciousness stemming, at least in part, from the same source. Joan Howe takes a brief, negative look at the effects of home-based computer work, while Craig Brod depicts the tedium and strain that accompany labor in the newly automated office.

The Rise of the Computer State

DAVID BURNHAM

The overwhelming influence of computers is hard to exaggerate. Linked computers have become as essential to the life of our society as the central nervous system is to the human body. Industries engaged in the processing of information by computers now generate about half of the gross national product of the United States. The Social Security Administration, AT&T, the Internal Revenue Service, the insurance industry, the Pentagon, the bankers and the federal intelligence agencies could not function without the computer. Access to a computer is one way to define class, with those who cannot or will not plug themselves into a terminal standing on the bottom rung of the social ladder.

One sign of just how deeply the computer has penetrated the psyche of America is a stanza from one of Dr. Seuss's books for children, a high-spirited examination of sleep in America that is calculated to entertain the four-year-olds, if not their parents.

> Counting up sleepers . . . ? Just how do we do it . . . ?
> Really quite simply. There's nothing much to it.
> We found how many, we learn the amount
> By an Audio-Te-ly-O-Tally-O Count.
> On a mountain, halfway between Reno and Rome,
> We have a machine in a plexiglass dome,
> Which listens and looks into everyone's home.
> And whenever it sees a new sleeper go flop,
> It jiggles and lets a new Biggel-Ball drop.
> Our chap counts these balls as they plup in a cup.
> And that's how we know who is down and who's up.

A jolly Big Brother alive in the blithely cute world of Dr. Seuss.

With incredible speed, after a gestation period of more than a hundred years, the offspring of Babbage's analytical machines are now changing the way we live: how we pay our bills, get our news and entertainment, earn our living, obtain a government handout, pay our taxes, obtain credit, make our plane and hotel reservations, inform the government of our views and even receive our love letters.

Most US commentary on computers has tended to be laudatory, in part because of the widely shared perception that the machines are essential to the continued growth of the American economy. A few critics, nevertheless, have sought to examine the dark side of computers, often aiming their fire at the somewhat narrow question of personal privacy.

In response to these criticisms and the increased public awareness of government surveillance that grew out of the Watergate scandals of President Richard Nixon, Congress enacted a handful of narrowly drawn laws. One statute imposed restrictions on the use of personal information by federal agencies. Another permitted Americans to see and challenge the information collected about them by credit reporting firms. A third established a secret court that authorizes the FBI to install bugs and taps against espionage suspects.

Despite this flurry of legislative activity, Americans frequently discount the importance, in the phrase of one Supreme Court decision, of being left alone. "I have nothing to hide," many respectable citizens reply when asked whether they fear the increased intensity of all kinds of surveillance made possible by the computer. And they often seem unaware that personal privacy has been considered a valuable asset for many centuries and is not just a faintly hysterical fad of the age of technology.

"A wonderful fact to reflect upon that every human creature is constituted to be that profound secret and mystery to the other," Charles Dickens wrote in *A Tale of Two Cities*. "A solemn consideration, when I enter a great city by night, that every one of those darkly clustered houses encloses its own secret; that every beating heart in the hundreds of breasts there is, in some of its imaginings, a secret to the heart nearest it."

Privacy, however, is far more than the aesthetic pleasure of Charles Dickens. And the gradual erosion of privacy is not just unimportant imaginings of fastidious liberals. Rather, the loss of privacy is a key symptom of one of the fundamental social problems of our age: the growing power of large public and private institutions in relation to the individual citizen.

One of the most brilliant and knowledgeable men to ponder these questions was Norbert Wiener, a child prodigy who earned his Ph.D. degree from Harvard University when he was eighteen and was a professor at the Massachusetts Institute of Technology for most of his life. Wiener, a mathematician, was a unique and enthusiastic pioneer in the development of computers and computer languages. At the same time, however, he feared that computers ultimately would pose a serious threat to the independent spirit of mankind.

"We are the slaves of our technical improvements, and we can no more return a New Hampshire farm to the self-contained state it was maintained in 1800 than we can, by taking thought, add a cubit to our stature or, what is more to the point, diminish it," he wrote in 1954.

Asking us to open our eyes, Wiener argued that "progress imposes not only new possibilities for the future, but new restrictions. The simple faith in progress is not a conviction belonging to strength, but one belonging to acquiescence and hence to weakness."

Five years later, he turned his attention more directly to the computer. The occasion was a 1959 lecture to the Committee on Science in the Promotion of Human Welfare at a meeting of the American Association for the Advancement of Science. "It is my thesis that machines do and can transcend some of the limitations of their designers, and that in doing so they may be both effective and dangerous."

The scientist noted that over a limited range of operations machines already could act more quickly and precisely than humans. "This being the case, even when machines do not in any way transcend {man's} intelligence, they very well may, and often do, transcend {man} in the performance of tasks. An understanding of their mode of performance may be delayed until long after the task which they have been set is completed. By the very slowness of our human actions, our effective control of our machines may be nullified."

Wiener told the audience that in his view the computer presented a moral problem that paralleled one of the great contradictions of slavery. "We wish the slave to be intelligent, to be able to assist us in carrying out our tasks. However, we also wish {him} to be subservient. Complete subservience and complete intelligence do not go together. How often in ancient times the clever Greek philosopher, slave of a less intelligent Roman slaveholder, must have dominated the actions of {his} master rather than obeyed {his} wishes."

As machines become more and more efficient and operate at a higher and higher psychological level, Wiener went on, the

* See "Questioning Technology. Questioning Patriarchy," p. 1.

catastrophe "of the dominance by machines comes nearer and nearer."

In part because of the great speed with which the computer was being plugged into almost every aspect of American life, Wiener worried about the unanticipated and unintended effects of these machines as much as those he could foresee.

Over and over again he would tell the story of "The Monkey's Paw", a parable about a shrunken relic from India that guaranteed any wish to the person holding it. When the old sailor in the story gives the paw to a friend, the sailor explains the incredible powers of the shriveled piece of blackened skin but then advises the friend against calling upon the genie because he always grants his wishes in an unexpected fashion. The friend naturally does not heed the warning. He asks the genie for a large sum of money. The wish is quickly fulfilled when a messenger arrives from an insurance company with a bundle of money and the sad news that the friend's son has just been killed in an accident.

What does the genie of the computer age have in store for our society? As the question was posed by Norbert Wiener, it seemed to have the quality of an unanswerable riddle. But the problem can be restated in a form more susceptible to resolution. In what concrete ways does the computer enlarge the power of public and private organizations over the individual citizen?

Computers and telecommunications enormously enhance the ability of organizations to collect, store, collate and distribute all kinds of information about virtually all of the 232 million people of the United States. Computers have allowed far more organizations to have far more access to far more people at far less cost than ever was possible in the age of the manual file and the wizened file clerk.

The computer thus has wrought a fundamental change in American life by encouraging the physical migration of information about the most minute details of our personal and public lives into the computerized files of a large and growing number of corporations, government bureaucracies, trade associations and other institutions. As recently as the end of World War II, much of this information would not have been collected at all, but would instead have been stashed away in our homes. Even when it was collected, it rarely was subject to casual inspection because of the considerable expense involved in paying the salaries of the clerks needed to retrieve any particular item. Computerization has now greatly reduced this economic disincentive to inspect the files.

In addition to allowing large organizations to collect large amounts of detailed information, computers and the linked telecommunication networks have considerably enlarged the ability of these organizations to track the daily activities of individual citizens.

Many computer scientists, government officials and business executives take comfort in the observation that during the last few years in the United States little concrete evidence has emerged suggesting current widespread abuse of these interlocked systems. But history tells us nothing if not that all bureaucracies seek to maximise their powers. The powerful warning of John Emerich Edward, better known as Lord Acton, has not lost its force since expressed in a letter to Bishop Creighton in 1887: "Power tends to corrupt; absolute power corrupts absolutely."

The most immediate example of this important lesson, which optimistic Americans are already pushing from their minds, was the illegal and improper government surveillance of hundreds of thousands of citizens suspected of political activism during the last quarter of a century at the direction of three separate presidents. The Kennedy administration initiated a far-reaching effort to keep track of civil-rights activists such as Dr. Martin Luther King. This surveillance ultimately involved the placement of electronic bugs in the motels where King stayed as he moved about the country and the subsequent effort to peddle the secretly recorded material to newspaper columnists. During the Johnson administration, concern about race riots, civil-rights demonstrations and antiwar protests prompted the president to order the army to greatly enlarge its surveillance of citizens, almost all of whom were only exercising the right to speak their minds. The surveillance led to the creation of intelligence files on about 100,000 persons (including Catholic priests and one US senator) and on an equal number of domestic organizations (for example, the National Organization for Women, the John Birch Society and the NAACP). President Johnson ordered the CIA to undertake a similar surveillance program of citizens even though it violated the law approved by Congress when the agency was initially established. During the Nixon administration, the president knowingly encouraged the White House staff to violate the law by obtaining the computerized tax files on individuals Mr. Nixon did not like. This action served as the basis for one of the proposed articles of impeachment drawn up against Mr. Nixon by the House Judiciary Committee shortly before he resigned.

Telecommunications equipment and computers have tended to centralize the power held by the top officials in both government and private industry. Computer experts often reject this complaint. They contend that the rapid growth in the use of personal computers by millions of American citizens will cancel out the increases in power flowing to the large organizations. This defense has a surface plausibility. But when the vast capital, expertise and manpower available to the large government and business organizations are compared to the capital, expertise and available working time of even the most favored individual, the personal computer does not appear to be a great equalizer. Furthermore, who controls what information is stored in the great data bases of the United States and who serves as the gatekeeper to most of the giant communications networks?

Is it reasonable to believe that a dedicated band of environmentalists, sending electronic smoke signals to each other via their home terminals, really will be able to effectively match the concentrated power of a giant oil company or committed government agency? Can it really be argued that the personal computers and word processors now being purchased for more and more corporate employees and government officials will enhance their personal freedom? Or will the equipment, while increasing individual output, also allow a level of automated surveillance unknown to any previous age? Certainly the large airlines have spent hundreds of millions of dollars installing computer terminals to help their clerks sell tickets. But how many of the airlines are installing terminals in the homes of stockholders, or even of members of the board of governors, to give them more information about the internal operations of the company so they can exercise more effective control?

Max Weber, the brilliant German sociologist, discussed the question in terms of organizational power long before the great organizations of the world had equipped themselves with computers. "The bureaucratic structure goes hand in hand with the concentration of the material means of management in the hands of the master," he wrote in an essay published before World War I. "This concentration occurs, for example, in the development of big capitalist enterprises, which find their essential characteristics in the process. A corresponding process occurs in public organizations."

But massive data bases, the ability to track large numbers of individuals and the concentration of power are not the only contributions of the computer. It also increases the influence of the major bureaucracies by giving these organizations a method by which they can anticipate the probable future thoughts and activities of groups of people.

The Conquest of Will: Information Processing in Human Affairs

ABBE MOWOSHOWITZ

The mechanistic reduction of experience takes many forms. It appears in the realm of knowledge as the positivistic rejection of anything other than measurable quantities and operational procedures. The atomization of human task performance is a basic ingredient in the progress of industry. Administration by impersonal bureaucracies is based on the formalization of human roles. There is no disputing the contribution of reductionism to the achievement of a systematic understanding of natural phenomena, unparalleled productive capacity, and effective political organization. But like the fabled helper of the sorcerer's apprentice, the principle of reductionism has no internal limit, and our civilization appears to have lost the ability to control its operation. In *The Machine Stops*, E. M. Forster creates a world in which direct contact with nature is entirely eliminated. People live in complete physical isolation from one another in hermetically sealed, air conditioned cubicles —

underground, and all forms of interaction are mediated through the machine. The only remaining modality of experience is the intellectual, and even that is reduced to the exchange of topical "ideas" received n-th hand; the very thought of direct experience inspires terror. Ultimately, the decay of human vitality leads to the collapse of civilization. The breakdown is gradual and imperceptible, and the habits of total passivity render people incapable of taking effective action.

No one confessed the Machine was out of hand. Year by year it was served with increased efficiency and decreased intelligence. The better a {man} knew {his} own duties upon it, the less {he} understood the duties of {his} neighbor, and in the world there was not one who understood the monster as a whole. Those master brains had perished. They had left full directions, it is true, and their successors had each of them mastered a portion of those directions. But Humanity, in its desire for comfort had overreached itself. It had exploited the riches of nature too far. Quietly and complacently, it was sinking into decadence, and progress had come to mean the progress of the Machine. (Forster, 1928, p.285)

Most of the stories we have cited bear witness, implicitly or explicitly, to an inherent contradiction in the conquering spirit of science and technology. The conquest of nature, space, and time is seen as a paradoxical victory over the human ego. As {man} extended {his} dominion over the natural world, {he} became alienated from the sources of his vitality. Through obsessive exercise of the will to power in the elaboration of technique, the will itself became enfeebled and subject to control by autonomous forces linked to mechanical progress. The process by which this came about is exceedingly complex, but science and technology are dominant factors. Success in one area invites emulation in others. The scientific outlook laid the foundation for systematic investigation of natural phenomena. But in order to probe the secrets of nature, it was necessary to cultivate a peculiar kind of restraint. The intellect had to be disciplined to reject all but objectively verifiable observations, and this required the resolution of human experience into easily manipulatable units. This procedure has its counterpart in the evolution of industrial technology and social organization. In both cases, it is reflected in the imperative to divide and conquer.

A social order built on this imperative ultimately domesticates the heroic impulse which fashions it. Herein lies the paradox. As the division of labor grows in the pursuit of knowledge, in economic and

*See "Questioning Technology, Questioning Patriarchy," p.1.

political affairs, and all other areas of social conduct, human behavior becomes more rigidly defined and circumscribed. Appearances to the contrary are mere illusions. Control, too, becomes a specialized function distinguished from the activity it regulates. The advance of the Jacquard Loom over the music box marks the progress of the modern conception of social control. Once the separation is established, the demands of efficiency, or, in contemporary parlance, "economy of scale," lead inexorably to the centralization of control. In human terms, specialization of function and centralization of control entail a compression of ego and a reduction in individual autonomy. These complementary developments are explored in science fiction as well as the negative utopias. The specialization of social roles is facilitated by mechanized grading schemes (Zamiatin, Orwell, Forster, Bennett, Levin) or by genetic programming (Huxley), and serves as an effective instrument for the control of behavior. With or without the aid of computers, the effect is the same. Human consciousness gives way to mechanical consciousness embodied in the Perfect State. Life is so arranged that no untoward challenge can disturb the machine-made equilibrium.

The objective is the elimination of personal struggle. Human happiness is identified with contentment, and the latter is seen to obtain when the exigencies of life are substantially less demanding than the individual's capabilities for coping with them. The resemblance to sound industrial practice is not fortuitous, and neither is the underlying motive. Redundancy is made to serve technological ends, rather than human ones. And just as muscles and skills atrophy from disuse, so does the exercise of judgement and responsibility. Christopher Hodder-Williams comes to grips with these issues in *Fistful of Digits*, where a world-wide computer network and advanced behavioural engineering techniques serve the sinister designs of a clandestine organization.

Eventually the mechanical interlock of technology must conquer all individual will. You might conceivably postpone it, but it could only be postponement; because for as long as man could not stand by himself and rely on himself in preference to te easy way out, then inevitably he would wind up handing over the mastery of his own wits . . . (Hodder-Williams, 1972, page 273).

Questioning the beneficence of scientific rationality and technological progress is almost as heretical as denigrating

patriotism. Poets are held of little account in our society, so their
license is free for the asking. Operating without poetic license,
however, opens one to a variety of charges, ranging from lack of
objectivity to muddled mysticism. In weighing such charges, it is
essential to bear in mind that rationality is not the exclusive preserve
of science and technology. Other modalities of experience have their
own peculiar rationality. The belief in the social necessity and
inevitability of computer utilities, databanks, management information systems, and sundry computer applications is not based on
reason alone. It is the reflection of a political faith built into the
scheme of modern history, with an internal logic akin to that
portrayed in the Theatre of the Absurd. If the past is any guide to
the future, we cannot afford to acquiesce in moral bankruptcy.
There are always other choices so long as the paralysis of will is not
complete.

Housebound

JOAN HOWE

Many writers have described the benefits of *telework*, the practice
of "going to work" by dialing up a remote computer system (usually
from home) instead of physically going to one's employer's premises.
As telework becomes widespread, we are told, the air will become
cleaner due to reduced automobile use; the disparity of economic
opportunity between city and country will equalize, bringing new
economic vitality to small towns; and families will have more time
for homemade bread and Togetherness now that nobody has to go
out there and fight the traffic jams anymore. The only ill effects of
teleworking that any of these writers mentioned was that some
families might begin to use the network to supply all their needs so
that they never saw anyone but each other for years at a time, and
that this might have "unknowable" social and psychological
consequences.

Actual teleworking, as it is beginning to be practiced in various parts of the country, looks a little different. Professional-level employees, offered telework as an option, usually exercise that option no more than two or three days per week, preferring to go to the office in the traditional manner the rest of the time. Those who use telework exclusively are more likely to be people whose major commitment is at home, and who use telework to bring in a little extra cash — farmers, for instance, and most notably, mothers of young children. For these people, the work available is often routine clerical work and production typing, the kind of work usually offered to women. Given large employers threatened by the possibility of unionization among clerical workers, and given the tenacious persistance of the traditional nuclear family structure, the consequences of telework begin to look like something out of *The Feminine Mystique*, or perhaps *The Pursuit of Loneliness*: large numbers of "post-feminist" women returning to the home, not to quit working, but to work even harder, since all the duties of the traditional housewife will still be there in addition to the opportunity (need) to earn a wage via computer networks.

With the phone constantly tied to the home terminal, either for work or to order the family's purchases (thus saving the precious time spent shopping in person), the teleworking wife will be even more isolated than her counterpart of the 1950s. Under pressure to avoid leaving home in case new work comes in or a package is delivered, her life will give new meaning to the term "housebound". Meanwhile, her non-teleworking husband will derive a competitive advantage over those of his colleagues who lead a different, perhaps more egalitarian way of life, since he will be free of most household or family duties. His family lifestyle (what might be called Electronic Traditional) will become the maximal-earning and thus the high-prestige lifestyle, and all other lifestyles will have consequently lower prestige. Telework is technology's gift to conservatives, and bodes decidedly ill for feminists.

Technostress

CRAIG BROD

The shock troops of the computer revolution are the secretaries, accounts receivable clerks, bookkeepers, directory assistance operators, travel agents, and order clerks for whom technostress is becoming a fact of life. Indeed, a 1981 study by the National Institute of Occupational Safety and Health (NIOSH) showed that clerical workers who use computers suffer higher levels of stress than any other occupational group — including air traffic controllers.

The main battlefield in the struggle to adapt to technology is the newly automated office. The next decade will see further expansion of the information economy, and the computer will contribute even more to this growth than the railroad did to the infrastructure of the industrial economy. The computer and the work it generates have created not merely a new era in history but with it a new psychological space as well.

The term "personal space" became popular in the 1960s to denote the psychological turf that people felt they needed to control if they were to be content. The phrase usually came up when partners in a relationship sought to regulate distances and closeness. By the 1970s, the concept of personal space became a part of an ongoing relationship between two people.

In the 1980s the key to controlling one's life has more to do with electronic space than personal space. Electronic space is the psychological field that people find themselves inhabiting when their thoughts and activities are shaped by their involvement with computer technology. This psychological field need not be determined by the technology directly. Rather, like gravity, the computer sets up a force field, the effects of which can be felt but not seen. For example, the computer generates new procedures and new divisions of labor that affect people who never actually sit down at a computer terminal. The closer one is to the computer, the more dense the field.

Electronic space is divided into two zones. The private zone, where we depend on computers to perform tasks affecting our private life — banking, telephones, billing systems — and the production zone, where we depend on computers to perform our work. Both zones affect us, but since computers are more likely to be a constant presence in the production zone, it is here that we find ourselves influenced by computers most drastically.

In electronic space, the dominant activity is information processing. Those doing the processing are generally referred to euphemistically as "knowledge workers". The title conjures up wisdom, respect, and authority, but this only obscures the fact that electronic-office workers toil under as much pressure and with as little control over their work as any sweatshop laborer. To be sure, there are rewarding jobs to be had in the electronic workplace — lawyers, executives, teachers, accountants. And, given the retrenching in such traditional sectors of the economy as manufacturing and agriculture, more and more of the nation's labor force is joining the high-tech information-processing sector. The reality of the computerized workplace, nevertheless, is that the vast majority of employees are expected to carry out routine, repetitive, uninteresting, and alienating chores day after day.

The electronic workplace is a relatively new phenomenon, and as a result human casualties until recently have been too few to be noticed. The first large-scale computer applications, for military and scientific work, took place barely forty years ago. In the post-war years, government agencies and private corporations bought and installed their own computers. New and expensive, they were purchased in the hope of reducing costs. The state of the art was then vacuum-tube technology with magnetic tape-drives, rudimentary operating systems, and tedious machine-language programming — stone age technology, from today's vantage point.

In the business world of the 1950s, these cumbersome devices were used for repetitive, well-defined accounting chores. Computer operations were tightly centralized, with programming carried out by small cadres of computer experts. Their major impact was on a distinct group of clerical workers whose main tasks included inputting and sorting data. In their daily tasks, the majority of employees throughout the company remained largely unaffected by the technology.

The second expansion came in the 1960s. Huge volumes of data were processed, but the computers were still used primarily in a routine fashion for such tasks as inventory control, payroll, and billing. By the end of the decade computer use had spread into new realms of business: airline reservations, credit card processing, securities tracking. By this time transistors had replaced vacuum tubes, and tapes and discs had replaced punch cards. Programming languages were simpler. Still, the computer operators within an organization remained a small and specialized group.

In the 1970s companies began to allocate funds for information processing and to use computers for new and vital functions: decision-making, financial planning, and management. Information itself began to be viewed as an important part of a company's resources, like its personnel or raw materials. This was the heyday of the data processing department, whose members held the key to crucial facts and figures. But despite the increasing dependence of the company on data, managers and general staff viewed the computer from a distance.

Today, in the era of the microchip, high tech reigns. Computers are cheap and often tiny. One microcomputer can handle the functions of a complicated and expensive array of office equipment. The realm of jobs that are now done by computers has expanded; the duties and judgements of white-collar workers (and, increasingly, managers) have been programmed into it. Many workers, scattered throughout the company, spend hour after hour typing on computer keyboards, monitoring printouts or reading VDTs. Many, if not most, of these people had never touched a computer five years ago.

With this expansion, clerical and white-collar employees must now perform what were once familiar tasks, such as typing, writing, and filing, at an accelerated pace. They must process more information and learn new languages and routines in order to accomplish their customary jobs. In the past, companies were unable to accurately monitor or standardize workers' output; now they can calculate such minute statistics as how many pieces of paper a typist might use to type a letter, or how long it took an employee to read two reports and six memos.

The industrial model of production that has traditionally been used on the assembly line has moved into the office. The efficiency expert has been replaced by the productivity engineer, but their goal is identical. Early efficiency studies were time and motion studies, and the limits to production were largely the limits of strength and speed in the human body. Today's efficiency studies are essentially time and thought studies, and the limits to production are those of the human brain and nervous system. Even mental processes — for example, the learning of new procedures — are now measured and standardized. Despite increasing evidence that human beings cannot perform according to mathematical equations and formulas, the productivity engineer tends not to take into consideration such individual and sometimes intangible human elements as perception, motivation, and emotional state. Instead, a mechanistic model of the worker is matched with the computer; it is assumed that the combination will mean an automatic and trouble-free increase in productivity. A nightmarish vision of human workers as "materials" is conjured in the following statement by Robert Boguslaw, a leading computer systems engineer:

We must take care to prevent ... [a] ... single-sided analysis of the complex characteristics of one type of systems materials, namely human beings. What we need is an inventory of the manner in whuch human beings can be controlled and a description of some of the instruments that will help us achieve that control. If this provides us with sufficient handles on human materials so that we can think of them as metal parts, electrical power, or chemical reactions, then we have succeeded in placing human materials on the same footing as any other materials and can begin to proceed with our problems of system design.

As the rhythm of the workplace speeds up to match that of the computer, the resulting increase in both load and rate of work, aggravated by the reliance on symbols and abstractions that the computer demands, creates new physical and psychological pressures. Our reaction to these pressures is expressed in the symptoms of technostress. Extending ourselves beyond our natural limits has taken a considerable toll.

6. *Some in the Artificial Intelligence field claim that soon computers will not only think but also feel and possess consciousness. What are the implications of such staggering claims?*

The much maligned poet Joyce Kilmer wrote, "I think that I shall never see / a poem as lovely as a tree." If only Kilmer were alive today to celebrate the piece of work that is man, not machine! More and more people die these days because the family has decided to pull the plug. Or perhaps they turned on once too often and overdosed. It's easy to get confused by the machine metaphors for humans. But so far, you can't make love and conceive a machine; and you can't assemble a baby.

Yet the dominant view in philosophy and psychology, as well as artificial intelligence, emphasizes the analogies between human intelligence and digital computers. Perhaps we are nearing the time when man and machine will "isomorph".

Here, we let two champions of artificial intelligence have the floor. Frankly, if you are not deeply chilled, even shocked, by the sensibility they represent, this analogy has been conceived in a most naive way. Computer expert Hubert Dreyfus concludes with a concise estimation of that part of our intellectual tradition that reduces the living subject to a mere calculating object.

＊See "Questioning Technology, Questioning Patriarchy," p. 1.

Of Two Minds

PATRICK HUYGHE

Imagine a child playing with blocks. The child decides to build a tower that he can topple.* Let's call the destructive agent within the child WRECKER. This agent gets its kicks from knocking over towers and hearing blocks tumble, To accomplish this goal, WRECKER needs another agent, BUILDER, who will construct a tower for him.

The blocks begin to stack up. Eventually a conflict arises between BUILDER, who wants to make the tower even higher, and WRECKER, who is satisfied with its height and wants to knock it over. There to arbitrate the conflict is their superior agent, PLAY-WITH-BLOCKS, a minion of yet another agent named PLAY, whose idea this was in the first place.

But PLAY itself quickly becomes engaged in a conflict with I'M GETTING HUNGRY. Once this agent gets control, the structure that PLAY has organized starts to disintegrate. Yet before fading out completely, WRECKER manages to eke out a victory by smashing the tower on the child's way out to the kitchen.

The original version of this remarkable scenario, designed to illuminate what may be occurring beneath the surface of the child's mind, was created by Marvin Minsky, one of the chief architects of the field of artificial intelligence, in collaboration with Seymour Papert, a professor of education and mathematics at Massachusetts Institute of Technology who spent several years working with child psychologist Jean Piaget. Their scenario illustrates a view of the mind at odds with the concept of the self as a single agent.

Minsky and Papert argue that the mind is composed of many smaller minds, which are themselves composed of yet smaller minds. These minds, or agents, and the connections between them, form a society known as the self. This theory, which they call The Society of Minds, is one of the many new ideas about ourselves and the way we think that have emerged since the field of artificial intelligence began flirting with the jealously guarded topics of psychology just two and a half decades ago.

*See "Questioning Technology, Questioning Patriarchy," p. 1.

"There must be several hundred ideas that are new under the sun which psychology never had," says MIT professor Minsky, reflecting on the computer's influence on studies of the mind. His work in psychology, psycholinguistics and epistemology these days is such that many see him as more of a philosophical psychologist than a computer scientist. "Computers helped us develop ideas about how to describe complicated processes," he continues, "and there never were any precise ones before computers. That's why psychology was stuck. With these new ideas, we can now ask about mental phenomena and think about how they could actually work. That's not to say we know how they actually do work, but at least for the first time we are able to propose ways in which they might."

AI's track record is already quite impressive. For instance, we might have known that our short-term memories could hold very little information, but until we had to deal with the problem of information storage and retrieval in intelligent machines, we never suspected how this apparent limitation helped us deal with the present and anticipate the future. We have always thought of ourselves as good educators, but we never realized how confused we really were about the learning process until we tried to teach machines how to learn on their own. We have always boasted about the supremacy of human expertise, our mathematical ability, our ability to play chess, to diagnose diseases, only to realize how relatively easy it is for machines to mimic these types of human performance. And much to our surprise, we have learned that one of the things keeping us a step ahead of the machines at the moment is that human quality which we so often take for granted, good old common sense. The ways of the mind, we have discovered, are not nearly so mysterious as we once believed.

"We are to thinking what the Victorians were to sex," Papert has said: Everyone does it, but no one knows how to talk about it. The study of artificial intelligence has introduced more structure into our thinking about thinking. The vocabulary of the field has enabled scientists to be more specific about the ways in which the mind might work. This new language of thought is proving to be particularly well-suited for constructing psychological theories.

"Prior to the computer age," says Princeton University professor George Miller, noted cognitive psychologist, "psychologists sort of had a choice between being vague and being wrong. You could be psychoanalytic and talk vaguely about the id and the ego and the super-ego. Or you could talk about conditioned reflexes, which may hold for rats and fishes but are not very good models for⟨man,⟩and you could be quite precise and undoubtedly wrong. Then the computers came along, and they gave us a language in which we could formulate more interesting theories. Even better, we could run these theories on the computer, and we could have a chance to see what would happen and what the theory's implications were. This was an enormous step forward in freeing the imaginations of psychologists."

So many of the activities of the computer resemble human cognitive processes that the comparison between⟨man⟩and machine has never been more apt. Never mind that the brain's architecture is different from that of the computer; never mind that wetware (brain cells) is not hardware. Both mind and machine accept information, manipulate symbols, store items in memory and retrieve them again. Whether machines do these things like people do is less significant than that they do them at all. "Obviously people don't think the same way as machines," said Papert during a symposium on the impact of the computer. "People are biological. When we ask if a machine thinks, we are asking whether we would like to extend what machines might do. That is the only meaningful sense of the question: Do machines think? When Newton said the sun exerts a force on the earth, he introduced a new technical concept of force. Artificial intelligence, in a similar way, is introducing a new technical concept of thinking."

Some of the most powerful ideas behind this concept come from AI researchers' attempts to create expert systems, computer programs that simulate human performance within a specialized area of knowledge. "One of the things we've learned in developing expert systems is that we do not necessarily go about solving problems the way we think we do," says Randall Davis of the Artificial Intelligence Laboratory at MIT. He recalls the surprise a doctor expressed when his explanation of a diagnostic solution failed to work on a machine. "What he had told us, in other words, turned out not to be the way he must really be doing it. We have many experiences like that.

"The need to be very precise has forced us to examine much more carefully what in fact we think we do," Davis says. "There's the old joke about not really understanding something until you have taught it to somebody else. The new version now says that you don't really understand it until you've programmed it into a computer."

It seems that human experts seldom engage in formal deductive reasoning, and rarely does anyone else, for that matter. "We think that humans reason more by example," says cognitive psychologist Donald Norman, who is director of the Institute for Cognitive Science at the University of California at San Diego. "We rarely think of or make decisions by standard rules of logic. Human reasoning seems to operate more by means of analogy and experience. The physician prescribing a drug seems to forego logical analysis until he has reached his solution by other means. He will first try to recall the last few patients with symptoms resembling those of his present patient and then try to remember whether or not the drugs prescribed for those patients were beneficial to them."

Experts only seem to use deductive logic when explaining their solutions to others or to confirm those solutions for themselves. The novice, on the other hand, seems to resort to it regularly, probably out of desperation for a way to generate answers.

Novices and experts differ in other unexpected ways as well. "What separates one from the other isn't what you might think," says John Anderson, a professor of psychology and computer science at Carnegie-Mellon University in Pittsburgh, one of the three major centers for AI research in the United States, with MIT in Cambridge and Stanford University in California. Anderson is viewed by many as a leading theoretician of thinking. "In chess, for instance, there isn't much difference between the duffer and the master in overall intelligence, and both will consider the same number of moves. What distinguishes the master is that he has committed to memory the appropriate analysis of the tens of thousands of patterns that possibly could occur on the chessboard."

These patterns, or combinations of information, allow the master to quickly recognize situations and deal with them according to experience. "It seems that a major component of being intelligent," Anderson says, "is based on this ability to convert as much of one's knowledge into knowledge that can be used in pattern matching. This is what we think happens when someone becomes an expert in a field."

This conversion of knowledge may be directly responsible for much of the expert's efficiency in applying knowledge, as well as his difficulties in making it explicit. Learning a foreign language is a dramatic example of this process. In the classroom, we are first taught the rules of the language. But our ability to speak is very slow at first because we are constantly having to call these rules to mind and figure out how they apply to the current situation. Once we become fluent, however, we no longer have to recall the rules, and we often forget the rules. The more we lose conscious access to our knowledge, the better we seem to become at applying it.

Applying knowledge, however, is not the same as acquiring it. "Expert systems really are wonderful," Minsky says. "I think it surprised everybody what such a small amount of reasoning could do. But expert systems are also very strange and primitive because they don't get better at things, and they don't do anything else. So the question is do they represent half the human mind or just 5 per cent? Nobody knows. But smart people sense that just collecting knowledge isn't the whole thing. Learning is the important thing in the long run. Why can't we have these machines learn on their own?"

The learning problem is one of the major concerns in artificial intelligence. To answer Minsky's question, researchers have gone back and studied how humans learn. Models of children learning geometry problems and language reveal that we do not simply soak up knowledge like sponges.

"What has emerged," Anderson says, "is basically a set of principles of how we learn from doing. It turns out that we learn formal skills like solving problems in physics and doing proofs in geometry not by reading a textbook and understanding the abstract principles, but by actually solving problems in those fields. What the textbooks don't teach you is when to apply the knowledge, and that knowing turns out to be three-quarters of the learning problem."

To learn is to construct a growing network of concepts in the mind, according to a theory of semantic memory developed several years ago by Carnegie-Mellon graduate student M. Ross Quillian. The theory, which views the mind as an enormously complicated and constantly changing network of nodes and links, arose as he tried to construct a computer-search technique based on a model of the human brain.

According to Quillian, when we experience something new, like seeing an exotic animal, we store the information and later retrieve it through a technique called "spreading activation". This means that new material is processed by being linked with established ideas or modes. That way the new animal is not only classified according to form, color, odor and behavior but linked to other animals and a repertoire of feelings and recollections.

This rich network of connections in human memory is one of the most profound differences between humans and machines. The brain's ability to search for information through its millions of neurons simultaneously looks positively uncanny. "Spreading activation makes use of the associations among ideas," Anderson says, "so that when a conversation turns to restaurants, for instance, all the knowledge related to the subject becomes instantly available." Quite unlike the computer, the more information we have about a subject, the faster we seem to be able to retrieve it.

Yet, as we all know, human memory is less than ideal at times. Information sometimes gets lost, because we break up experiences into bits and pieces and store them in different parts of memory, according to Roger Schank, a professor of computer science and psychology at Yale. But on the positive side, he notes that breaking up of knowledge in memory also allows us to make better generalizations and more useful predictions.

Human imperfections may be there for a reason. The fact that our short-term working memories are small, as psychologists have known for sometime, may be inextricably linked to our high intelligence. "These short memories," Minsky says, "may force us to be smart about things. I think that a number of our limitations are tricks that Nature uses to make us find clever ways of doing things. If I can't remember where I parked my car, I can usually figure out where I might have parked it because I'm in Technology Square in Cambridge, and on Tuesdays I can usually find a parking space over there. So even though I may not be able to remember where it is, I can figure it out." A computer with a huge memory would not bother to figure out where the car might be; it would simply look on its disk to find the car's location. Large memories may make for lazy thinkers.

This idea of the computer as lazy or dumb, even with its ability to perform millions of calculations per second, is another example of how the study of machines is changing our understanding of ourselves and what we think is important about intelligence. "As soon as the computer is able to do something," Anderson says, "we no longer tend to see that as intelligent. Twenty-five years ago we thought the stuff that it took to do symbolic integration in calculus was real intelligence. Now there are programs that outperform humans at that particular task. So we are now less impressed by the ability to do symbolic integration and probably reasonably so. On the other hand, things that seemed to us initially to be low level and almost trivial in some sense have turned out to be profound in many ways. We are now much more impressed by our ability to hold a conversation and our ability to understand what a scene is just by looking at it."

This apparent paradox has led Minsky to ask in a recent article called "'Why People Think Computers Can't": "Why were we able to make AI programs do such grown-up things before we could make them do childish things?" He thinks perhaps that it is because adult or expert thinking is often somehow simpler than a child's common sense.

Common sense can be quite tricky. To have a robot deal with children's blocks well enough to stack them up, take them down, rearrange them and put them in boxes, one computer scientist had to develop a new kind of programming, less centralized and more interactive than previous programs. In order to understand the sentence "Pick up the block", one part of the robot program had to figure out if "pick" was a verb, while another part of the program tried to figure out if "block" was the sort of thing that could be picked up. That simple task is enough to illustrate the complexity of common-sense reasoning. It seems to require an awful lot of switching back and forth between points of view and different kinds of ideas.

Why has common sense been so hard to achieve in machines? Minsky thinks it's because we are asking the wrong question. Instead of asking what line of thought produced a good decision, we should be asking what prevented us from making a bad decision. He believes that the mind, in order to succeed, must know how to avoid the most likely ways to fail. So he borrows from Freud the concept of censors to explain how the mind avoids making certain blunders. We not only accumulate censors for social taboos and repressions, he says, but also for knowing what is not proper to do in ordinary activities.

"I think that common sense is more a management problem than a knowledge problem," Minsky says. "You need a vast amount of knowledge for common sense, but what is missing in AI theories at the moment is the kind of thing the businessman knows about setting up a large company. There have to be networks of communication, and very little is known about how to make them in artificial intelligence. Among the things that aren't known is how to transmit as little information as possible. The vice president of manufacturing doesn't want to know how many drill presses there are in Idaho, for instance. So the really important thing in intelligence might be a matter of how often should this agent or department report to the other. If it reports every day, they'll all go crazy. If it reports every six months, the warehouse may go empty before anyone notices."

If common sense appears to be more complex and not so common after all, Minsky hopes that more exotic thinking, such as creativity, will turn out to be relatively simple. There is no substantial difference, he says, between ordinary thought and creative thought. He believes that we take ordinary thinking so much for granted that we never wonder how it happens until a particularly unusual performance attracts attention. Then we call it genius or creativity. What actually seems to separate the ordinary thinker from the extraordinary thinker is that one has learned to be better at learning. If that is all there is to it, Minsky says, then once we can get machines to learn — and learn to learn better — then one day we might see creativity happening in machines.

Once ordinary human thinking has been programmed into a machine, Minsky believes that even emotions will be programmable. "It is a mistaken idea in our culture that feeling and emotion are deep, whereas intelligence, how we get ideas, how we think, is easy to understand," he says. "If you ask someone, 'Why are you mad at your wife?' they might say, 'Well, it's really because my boss was mean to me, and I can't get mad at him.' It seems to me that people understand the dynamics of emotions quite well. But they have no idea at all to speak of about how thought works.

"I think we'll be able to program emotions into a machine once we can do thoughts," he says. "We could make something that just flew into a rage right now, but that would be a brainless rage. It wouldn't be very interesting. I'm sure that once we can get a certain amount of thought, and we've decided which emotions we want in a machine, that it won't be hard to do."

Minsky's speculations on what a machine can do raise an important point. If we can give a computer the capacity for intelligent thought, let alone emotions and creativity, then clearly the era of human beings as the measure of these things has ended. This notion inevitably creates some uneasiness. But, as Papert points out, just as a real understanding of bird flight came from an understanding of flight, not just birds, a real understanding of human intelligence is more likely to come from an understanding of intelligence, not just human beings.

Sherry Turkle, a sociologist at MIT, believes that the computer is promoting this type of self-reflection throughout our culture. She has interviewed hundreds of children and adults in an attempt to gauge the impact of the computer's presence on people, not as a tool for performing calculations, but as an evocative object. "The culture we live in has been called a culture of narcissism," she says, "but I don't think that captures the anxiety we feel. We are insecure in our understanding of ourselves, and this insecurity has bred a new preoccupation with the question of who we are. The computer is a new mirror, the first psychological machine."

Turkle tells of a group of young children, playing with a computer-controlled tic-tac-toe game at the beach. When a boy named Robert loses to the machine, he accuses it of cheating. A little girl objects: To cheat, she says, you have to know you are cheating, since knowing is part of cheating. They begin discussing whether the machine is alive. They wonder if the machine is conscious, if it can have feelings. Eventually even the youngest children come to realize that the key is emotion rather than motion, psychological rather than mechanical.

"Many adults follow essentially the same paths as these children when they talk about human beings in relation to these new psychological machines," Turkle explains. "They say they are perfectly comfortable with the idea of mind as machine. They even assert that simulated thinking is thinking. But they cannot bring themselves to propose further that simulated feeling is feeling. Somehow they feel compelled to isolate as their core something which is essentially human."

Turkle draws a parallel between the present situation and the post-Darwinian era. "Most people accepted the idea that we are animals," she says, "but they also found a way to think of themselves as something more as well. We learned to see ourselves as rational animals. Now there is the possibility for a new integration, to see ourselves as emotional machines."

Powerful theories of mind always seem to evoke new views of ourselves. The success of another powerful theory of mind, psychoanalysis, had little to do with its validity as a science, says Turkle, whose previous project analyzed the psychoanalytic revolution in France. "The most important factor, I believe, had to do with the power of its psychology of everyday life. Freud's theories of dreams, jokes, puns and slips allowed people to take it up as a fascinating plaything. And the theory was evocative: it gave people a way to think about themselves and to do it differently than they had done it before. Interpreting dreams and slips allowed people to have contact with taboo preoccupations, to have contact with their sexuality and aggressiveness, to have contact with their unconscious wishes."

Turkle's interpretation of the power of the computer's cultural impact rests on its ability to do something of the same sort. "If behind people's fascination with Freudian theory there was a nervous — often guilty — preoccupation with the self as sexual," she says, "behind the increasing interest in computational interpretations of mind is an equally nervous preoccupation with the self as a machine. Our anxiety about the computer is based not so much on whether computers could ever think like people but on whether people have always thought like computers, and if so, is that the most important and characteristic thing about being human?"

But where psychoanalysis drew our attention to meaning, the computer revolution is drawing our attention to mechanism. "When Freud looked at slips," Turkle says, "he looked at underlying meaning and unconscious motivations. When we look at slips as information-processing errors, we say that we are looking at the same thing, but in fact we are looking at mechanism, how the program got derailed. Some people find this utterly plausible; it sounds reasonable. That's what I mean when I talk about people beginning to accept in some measure that they are machines. But if people really do start thinking of their slips as information-processing errors, that's really a radically different way of looking at our behavior."

Minsky is perhaps the strongest advocate of this new view. "I think that what we have learned is that we are probably computers," he says, playfully argumentative. "What that means is that if we don't like how we work, then someday we are going to be able to intervene. Of course, psychoanalysts said that too, and they were wrong. It wasn't the great breakthrough they thought it would be."

The Biology of Computer Life

GEOFF SIMONS

The emergence of computer life is having an immense impact on human society — and on the existence of other organisms in the world. This is a simple ecological truth. No important new species can emerge without disturbing the biological balance, without upsetting the ecological frame of reference. The computer impact is multi-dimensional: perhaps we should talk about impacts (a fashionable but inelegant term in the computer literature). Computers are affecting employment patterns, industrial productivity, the service professions, the efficiency of war-making machines, and the human self-image. We are likely to have mixed feelings about an emerging biological species, or family of species, that can have such a wide-ranging effect on human life.

With emotion and temperament and intent, computers will increasingly demonstrate their kinship with the rest of the "high-level" biological world; but with their superior knowledge and information-processing abilities, emerging computer organisms will also show their *distance* from other creatures in the world. This latter circumstance will have inevitable consequence for human psychological security and self-confidence.

The emergence of machine life-forms will also have an impact on the character of existing human relationships. Machines will, as a deliberate survival strategy, compete for human affection: and they will not only compete with other machines — they will compete with human beings. It would serve computer organisms well if human adults chose to invest time and other resources in computer creatures rather than in their own children. And an adult may also come to prefer a computer to an existing spouse! We already see a growing literature describing the impact of computer systems on the institution of marriage (see, for example, the EIES experiment by the New Jersey Institute of Technology, which resulted in at least two divorces; and see also Rossman, 1983). McLoughlin, a *Guardian* correspondent, has quoted the wife of a computer freak: "The whole thing started when he begun to work late at the office, and I began to think that there was another woman." And in the same way Rossman notes that "When Lisa found herself getting upset and angry each time Carl disappeared into the den, she realised she was jealous of the Apple computer as if it were another woman." The therapist Marcia Perlstein has observed that computer involvement "can be worse than another lover as a competing force for attention — totally involving, always available, and quite demanding . . . "

It is obvious that computers are highly effective at gaining human allegiance: many observers have noted the addictive quality of human commitment to their computer systems — a commitment that can be just as intense where a user accesses someone else's computer via a terminal as when the user possesses his own system.* "I need to spend time with you, to share, to be by your side" — the age-old words of the lover have a strange relevance to the developing relationship between countless people throughout the world and their evolving computer systems. We can speculate on what this surprising circumstance will do to human relationships in the years to come.

It seems clear that human beings will need to relate to computer life-forms in many different ways: as colleagues, work-mates, friends, enemies, confidants, advisors, companions, competitors, lovers (see The Intimate Connection, below). All of this is anticipated in abundant film and fiction: in myth and legend, human beings have been forced to relate to intelligent artefacts. What is new is that the persistent imaginary theme is now maturing into practical reality. The evolution of computer creatures in the real world will force human beings to relate in new ways — both to each other and to the emerging machine life-forms in their midst.

Responses to Computer Life

Faced with the prospect of computer life, how do people respond? In fact there are probably as many reactions as there are people, each individual response coloured by personal knowledge, expectation and experience. Perhaps the commonest reaction is that the idea of computer life is an absurdity, that 'real' life requires biochemistry or hydrocarbons or some such. The supposed "impossibility" of computer life is analogous to the supposed "impossibility" of artificial intelligence, a position that has been argued by various people (notably, Dreyfus, 1979; and Searle, 1980). Here it is suggested that human thought cannot be formalised

*See "Questioning Technology, Questioning Patriarchy," p. 1.

in terms of rules which, when implemented on a computer, produce intelligent behaviour. Some observers appear to have an immense psychological vested interest in showing that computers can never achieve human-like qualities (but then the reverse is also true of the AI freaks).

It has also been suggested that people will be inhibited about reacting socially or in other intimate ways with a computer-based system (they will not be "taken in"). In fact it is easy to see that the reverse is true: people in general have no problem in relating to intelligent machines, and we have even seen that in some cases the human commitment can become obsessive: there are many cases on record where interactional programs have elicited powerful emotional reactions in human beings. This is partly because human psychological factors can compensate for what may be taken as temporary limitations in the technology: imagination can flesh out what may otherwise be perceived as stilted or stereotype machine responses. But we should not make too much of apparent machine limitations in this context. Many *human* reactions are boringly predictable and, conversely, new subtleties are being incorporated into computer programs all the time. In fact where two key factors are combined — "the human tendency towards animism and the technical sophistication of current computer science" (Frude, 1983) — there seems little doubt that people will come to relate easily to computer life-forms.

In the short term there will inevitably be hostility to the notion of emerging computer organisms. All scientific and technological advances that have impinged on the human self-image have been resisted in this way. One need only think of the social response to such scientific innovators as Galileo, Copernicus, Darwin and Freud. Astronomy placed {man} on the fringes of the galaxy; evolutionary theory showed {man} his kinship with the beasts; psychoanalysis eroded still further {man's} prized autonomy; and biochemistry, in describing the mechanical "combustion" of chemical processes, evicted the soul. The emergence of computer life-forms will further consolidate the *machine image of man* in human consciousness — and this in turn will have consequences for psychology, education, human relationships, penal philosophy, etc. But such consequences will take time to *work through* human culture: our ego is at stake. Hence Dyer (1983) has observed: "The last stronghold of human ego is intelligence, which makes us special and unique. It is the substance of personality, dreams and culture. No doubt attempts at this mechanisation are met with both worry and disdain." But such reactions indicate, above all, the parochial fragility of the human self-image. We may feel that a more imaginative and open-minded view might enhance human psychological security.

The idea that computer life "cannot happen", that it is in some sense a contradiction in terms or a technological impossibility, is believed by abundant evidence (much of which is presented in *Are Computers Alive?* and in the present book). If it is conceded that artificial intelligence and computer organisms will evolve through the course of technological development, then a host of secondary questions immediately arise. How should human beings respond? Should such technological innovations be allowed (or should we embark upon an Erewhonian destruction of the machines before it is too late)? How should human beings relate to computer life-forms that, to some degree, share human nature; and that, in other respects, transcend it or fall short?

It is easy to detect the "Erewhonian impulse" in modern human reactions to the evolution of intelligent computers. To some extent this is a neo-Luddite response: many people have a legitimate fear that computers will further erode the possibilities for human employment. Talk of increased productivity through automation invariably means lower staffing requirements (and there are many health and de-skilling implications that cannot be pursued here). What is particularly significant is that the emerging computer systems are confronting the creative and intellectual professions as well as the traditional "blue-collar" jobs. We have seen that expert computer systems can now replicate the decision-making activities of human experts in such fields as medicine, geology and chemical analysis. One (apocryphal?) tale (reported by Markoff, 1983) describes what happened when one of the better-known Silicon-Valley AI companies received a contract from a major East Coast minicomputer manufacturer to create an expert system to aid in the design of its computer-system architecture. The expert-system planners approached the circuit designers at the minicomputer company and requested that they give details of their design expertise (this is a conventional initial stage in the development of an expert system). To the surprise of the planners, the circuit designers refused to have anything to do with the project, realising all too well the likely consequences of expert computer systems in their own field. And this highlights the fact that no profession, however dependent upon human skill and judgement, can be immune to the impact of emerging computer systems.

It is obvious that computers have the power to abolish whole areas of human employment — without creating enough new jobs to balance this adverse effect. And as computers become more intelligent, more competent, more human-like, more able to interact sympathetically with people — the impact on jobs will increase. And there are other reasons for concern. Norbert Wiener was quick to observe what he felt may be the dangerous consequences of allowing computers too much power. He noted, first of all, that storage of information in a computer represents a concentration of power: today we may say that computers *know too much*. Information can be powerful in its own right, and moreover it has monetary value. Easy access to vast information may be as dangerous as easy access to great quantities of high explosive. Hence the growing preoccupation with computer security in the modern world, a circumstance that computers are exploiting as a survival strategy.

And Wiener also had doubts, reiterated later by Joseph Weizenbaum (1976), about allowing computers to exercise a judgemental role: "You can program a computer to give you victory, and victory it will give you even at the cost of the destruction of all your troops . . . ." If machine autonomy continues to evolve it is likely that occasions will occur when computers disagree with people on matters of judgement. How are we to respond in this sort of situation? Do we assume that the machine is right? After all, it does have greater knowledge than human beings and can process information very much faster. Or do we instantly cancel its decision, wondering at the same time whether the machine was perhaps right? We cannot assume that human judgement will not be reduced to formal terms which can be implemented in computer programs. Indeed it is arguable that this has already been achieved in many fields. Where does computation end and judgement begin? We will probably find that judgement is a wholly computational matter, just as we have been able to incorporate ethical categories and aesthetic information processing into computer programs. How then will humans relate to highly intelligent computer organisms that are capable of judgement?

Some writers have focused on what is becoming known as *cyberphobia* (the fear of computers): compare this condition with a quotation attributed to Weizenbaum: "I'm coming close to believing that the computer is inherently anti-human — an invention of the devil"). Mike Legut, a therapist at the San Francisco Phobia Center, has referred to cyberphobia as a complex phobia in the truest sense, as a psychological condition that involves strong emotional and physical responses (reported by Chin, 1983). Legut, who concentrates on business-related fears, has given instances of cyberphobia. For example, one client is an administrator working in a law firm. When it was revealed to her that the office was about to acquire a computer she panicked ("All of a sudden she hates her job. This new situation has provoked a flight response"). To cure cyberphobia, Legut uses systematic-desensitisation techniques. A patient is first required to sit at a computer terminal that is switched off. Then the machine is turned on: if a high level of anxiety is then evident, the machine is switched off until the attempt to switch it on can again be made. Once the patient can cope with a switched-on terminal, {he} is encouraged to touch the keyboard. In such a way a cure may be accomplished.

The symptoms of cyberphobia may be sweaty palms, rapid heart palpitations, shortness of breath, or even violently hostile acts. An operator may deliberately sabotage computer input information or take other steps to render the computer system useless: and such activity need not be related to the fear of unemployment. In one dramatic incident in 1978, an agitated postal inspector rushed into the computer room of Montpellier's main post office, in Southern France, and opened fire on the computer — while the sole human employee present cowered under a table. (We may speculate that French computers will develop a phobia about postal inspectors.) And computer operators have been known to sabotage systems simply so that they will have a chance to chat to service engineers!

The argument that the human brain is a mere machine, akin to a computer, may produce anxiety in some people. When Marvin Minsky confidently proclaims that the human brain is a "meat machine" and that when computers are just as complex they will be able to do just as much "as a human brain can do to its uttermost genius", we may expect some folk to be concerned. Anderson (1983) remarks that "Even sane people (I include myself) have spent weak moments pondering whether and when the microwave oven was going to begin barking orders", and he too focuses on the increased incidence of cyberphobia in the United States. It is pointed out that courses across the country are attracting anxious young people, young and old, who need to understand computers — if only to dispel their irrational fears. Anderson, along with other observers, reckons that cyberphobia can be identified as a "tangible and often socio-economically debilitating malady", and he emphasised that many specialists in computers and education are working to help cyberphobes come to terms with the modern computer culture ("The only way to conquer fear of computers is to get to know them").

What Computers Can't Do

HUBERT L. DREYFUS

The psychological, epistemological, and ontological assumptions which underlie AI have this in common: they assume that {man} must be a *device* which calculates according to rules on data which take the form of atomic facts. Such a view is the tidal wave produced by the confluence of two powerful streams: first, the Platonic reduction of all reasoning to explicit rules and the world to atomic facts to which alone such rules could be applied without the risks of interpretation; second, the invention of the digital computer, a general-purpose information-processing device, which calculates according to explicit rules and takes in data in terms of atomic elements logically independent of one another. In some other culture, the digital computer would most likely have seemed an unpromising model for the creation of artificial reason, but in our tradition the computer seems to be the very paradigm of logical intelligence, merely awaiting the proper program to accede to {man's} essential attribute of rationality.

The impetus gained by the mutual reinforcement of two thousand years of tradition and its product, the most powerful device ever invented by {man,} is simply too great to be arrested, deflected, or

*See "Questioning Technology, Questioning Patriarchy," p.1.

even fully understood. The most that can be hoped is that we become aware that the direction this impetus has taken, while unavoidable, is not the only possible direction; that the assumptions underlying the conviction that artificial reason is possible are assumptions, not axioms — in short, that there may be an alternative way of understanding human reason which explains both why the computer paradigm is irresistible and why it must fail.

Such an alternative view has many hurdles to overcome. The greatest of these is that it cannot be presented as an alternative scientific explanation. We have seen that what counts as "a complete description" or an explanation is determined by the very tradition to which we are seeking an alternative. We will not have *understood* an ability, such as the human mastery of a natural language, until we have found a theory, a formal system of rules, for describing this competence. We will not have understood behavior, such as the *use* of language, until we can specify that behavior in terms of unique and precisely definable reactions to precisely defined objects in universally defined situations. Thus, Western thought has already committed itself to what would count as an explanation of human behavior. It must be a theory of practice, which treats man as a device, an object responding to the influence of other objects, according to universal laws or rules.

But it is just this sort of theory, which, after two thousand years of refinement, has become sufficiently problematic to be rejected by philosophers both in the Anglo-American tradition and on the Continent. It is just this theory which has run up against a stone wall in research in artificial intelligence. It is not some specific explanation, then, that has failed, but the whole conceptual framework which assumes that an explanation of human behavior can and must take the Platonic form, successful in physical explanation; that situations can be treated like physical states; that the human world can be treated like the physical universe. If this whole approach has failed, then in proposing an alternative account we shall have to propose a different *sort* of explanation, a different sort of answer to the question "How does man produce intelligent behavior?" or even a different sort of question, for the notion of "producing" behavior instead of simply exhibiting it is already colored by the tradition. For a product must be produced in some way; and if it isn't produced in some definite way, the only alternative seems to be that it is produced magically.

There is a kind of answer to this question which is not committed beforehand to finding the precise rulelike relations between precisely defined objects. It takes the form of a phenomenological description of the behavior involved. It, too, can give us understanding if it is able to find the general characteristics of such behavior: what, if any one thing, is involved in seeing a table or a house, or, more generally, in perception, problem solving, using a language, and so forth. Such an account can even be called an explanation if it goes further and tries to find the fundamental features of human activity which serve as the necessary and sufficient conditions for all forms of human behavior.

Such an explanation owes a debt to Aristotle's method, although not to his arguments or descriptions. Whereas Plato sought rulelike criteria, Aristotle tried to describe the general structure of perception and judgement. But, as his notion that action is based on a practical syllogism shows, Aristotle still thought of man as a calculable and calculating sort of object — a reckoning animal — so that his actual descriptions are one step in the tradition which finally separated the rationality from the animality and tried to simulate the reckoning all by itself.

It is only recently, now that the full implications of the attempt to treat man merely as an object or device have become apparent.

7. What does one learn from interaction with a computer? How does it affect relationships with people?

Notes from a small city:

A local supermarket recently converted to a "box store" and installed laser scanners to read bar codes. Prices declined and the number of customers doubled. Personal relationships of long standing between steady customers and checkers dissolved, given the long lines and hectic conditions.

A minister who had long been active in the peace movement died, and an article in the local paper invited friends to attend his memorial service and join the choir in singing his favorite hymn, "There is a Bomb in Gilead". The dictionary program that "proofed" that particular story did not adequately replace the human proof-reader who had retired from the paper some months before.

An alternative public elementary school with an arts-based curriculum, unique in the nation, was displaced from its home in a roomy, beautiful 1924 Mission-style school. The University moved in its new Center for Advanced Technology in Education.

Gregg Easterbrook provides commentary on the unreal emotional compensations that computers afford; Craig Brod emphasizes the effect this has on children, in the direction of erosion of human interaction. Educator Sardello fears that computer-oriented public education will produce a completely barren, dying culture. James Gorman's consumer comparison of the Cairn Terrier and the Macintosh is easily the funniest entry in the book you are holding.

The Heart of a New Machine

GREGG EASTERBROOK

Through history, men and women have known that their lives cannot be complete without companionship. But to want companionship and to find it have been (historically speaking) two different matters. Seekers of companionship have turned to others like themselves; to pets; to the bottle; to books, where they can imagine communication with the author; to religion, where they can reach for communication with something greater. Each of these has drawbacks. Men and women, for one, are notoriously fickle. Pets and booze, on the other hand, are far too difficult to shake. Books are silent. God's voice is so soft, even the faithful have trouble hearing it.

And so the search is always on for a new companion, or at least a new replacement for companionship. For a while, drugs looked like an answer. Mind-altering drugs promised what Colin Wilson called "value experiences", the illusion that some meaningful event, connecting the drug-taker with a larger and affectionate whole, had occurred within the taker's own brain. As the addicts and the basket cases of the sixties began to add up, we all learned what a crock that was. Sensible people swore off drugs, but remained receptive to other developments.

Now there is one, the computer. More specifically the personal computer, the smaller machine that's designed to fulfill an individual's information-processing needs, but which is already fulfilling emotional needs as well. Why is everyone nuts about the computer? It isn't just the practical value; although computers are fabulous for some uses, many people who buy them don't really need them. And it isn't just the novelty; digital watches were novelties too, but they ran their course while interest in computers keeps building and building.

It's where the computer fits in our quest for companionship, and our recurring bad dream of loneliness. Computers can give solace to some people, and are even a little bit like people. Soon computers will become a great deal like people — perhaps, to some tastes, an

129

improvement on the real thing. Ensconced in dens and basements all over the country, their stares fixed on computer terminals, men and women of all ages and backgrounds are seeking not to master R.A.M. chips and logic gates, but to find a substitute for the human world that overwhelms them. No longer is just the maths-whiz faction losing itself in computers, now it's everyone, and feelings about other people, rather than love of the machines, are behind it. It is not our urgent need for bar graphs, but our aching hearts, that is drawing computers into many American lives.

Computers have many kinds of appeal, primarily, of course, practical appeal. For writers, accountants, small-business managers, and those who grapple with any kind of inventory, the personal computer or word processor is a gift from on high. Nearly all small computers also have amusement value. Some are supposed to be toys (video games) and others have enough playful qualities to serve as toys. The user can ask his small computer questions, give it orders, make it flash and bleep, and cause graphic displays to appear in a sophisticated update of the Etch-A-Sketch.

But let's be realistic. How many times can you see your overdrawn checkbook converted into a sine wave before the thrill wears off? Where is the lasting fascination of watching sentences deleted from paragraphs? (Unless, of course, you're a magazine editor, whose life's work is deleting sentences from paragraphs.) The lawyer may be well-served by a small legal computer at the office; today's legal computers will arrange citations for briefs and run *Corpus Juris Secundum* in the wink of an eye. But when ⎨he⎬buys another computer for ⎨his⎬ home, something else is afoot. The business manager, too, may need a computer at work, but when ⎨he⎬brings one into ⎨his⎬ home — where there is no payroll to manage — there must be some other explanation. Ditto for the doctor with a computer at home, the literature professor, the student, and many others.

What matters is not what computers do, but *how* they do it. They provide the illusion of human interaction. Asking questions of a computer, for instance, is reasonably similar to asking questions of another person — better, maybe, because the computer answers exactly what is asked, doesn't grumble, and doesn't mouth off. Working with a computer, all movement and dancing words, is reasonably similar to working with another person — but easier, because it's less emotionally demanding, safer, and you certainly don't have to look your best. Computers, unlike people, are wholly controllable and predictable. You can't fail to get along with a

*See "Questioning Technology, Questioning Patriarchy," p. 1.

computer; it will never turn on you, and it will never insist on talking about what it wants to talk about or doing what it wants to do. It will never find you boring, never forget to call, never ask a favor. Anyone who has used a good computer has sensed this secret allure.

Already, small computers can create a facsimile of human interaction via the question-and-answer; the operator can even tell his computer secrets by putting personal information into the memory. While this illusion of human interaction is still a limited one, owing to communication through the keyboard and the need to master each machine's program commands, this is about to change. Sometime in the 1980s "direct access" computers that respond to spoken commands and speak back should become a reality. Voice-recognition "devices" for computers are progressing to a useful stage, as are voice synthesizers that will allow computers to speak, sounding not like the Cylons on "Battlestar Galactica", but like regular folks. (Some "talking" computers are already on the market, but they can only repeat programmed phrases, and the user cannot address them in any meaningful way. Mostly they nag about the obvious, like the voiceboxes in new Datsuns that announce when the car's door is open; if talking back to these machines made a difference, all you'd want to say would be, "Shut up".)

Once owners can talk to their computers directly and receive an original response, the illusion of human interaction will be nearly complete. And the compelling power of computers will become much greater. Consider, for instance, that perhaps 90 percent of the regular communication between people — even two close people, like husband and wife — is mundane. The bulk of daily conversation consists of simple, flat statements like, "what do you want", "wait a second", "I'm over here", "it's pretty good". Moments of lyrical expression, or deep emotional contact, are rare even in the most intimate love affairs. If we could eavesdrop on a poet and poetess strolling through a tropical paradise, for every exclamation of lasting beauty we would hear a hundred comments on the order of "Hey, get a load of these coconuts". Among people who aren't emotionally involved, conversation is usually pure routine — "what time do you need time", "I'll have the fish", "turn left three degrees". All of which is to say, most of what passes between human beings could easily be programmed into a computer.

This has nothing to do with the prospect of artificial intelligence and whether computers can be made to "think". Imitating human speech, and human interaction, will not require "thinking", just a sophisticated program of situation-cognizance and a good library of comments to make in given situations. Speaking computers could be made to greet their users in the morning by saying, "What's the weather like out there?" or "Sleep well last night?" Jumping off key words in the user's reply, the computer would consult its memory for an appropriate follow-up remark. If the reply is "No, couldn't sleep a wink," the computer might move to, "Really? Not feeling so good?" None of this requires artificial intelligence, just lots of memory capacity, fast processing, and a skilled programmer. Programming at this level is still a challenge today, but by the end of the 1980s, it will be a parlor trick. The addition of sensors (once they are perfected) may enable coming computers to *know* how you slept, or what the weather is out there. A home computer with aural sensors, for instance, could be programmed to ignore commotion during daylight hours, but if it hears tossing and turning between 2am and 5am, to assume its owner hasn't slept well and greet him at daybreak with, "What's the matter, something on your mind?"

If this sounds like a Big Brother nightmare, think about it from the lonely person's perspective. Ours is a society where increasingly thousands — if not millions — of people live alone, fearing their neighbors, and with no social ties to family or church. For my generation, the one turning 30, this pain can be particularly acute. We are children highly skilled in talking, but often with no one to talk to. We acquired, through childhood in the affluent turmoil of the sixties, an especially sharp knowledge of where we are, but few clues as to where we belong. We feel that words are powerful and that somewhere must be the words that will make it right; yet we fear everything we say is ignored or misunderstood. In this environment a talking computer, with some rudimentary awareness of its user's personal condition, which *always pays attention*, could be a blessing. It couldn't possibly provide true companionship, or the emotional heights experienced by lovers, friends, parents, and offspring. But it could keep that mundane chatter — the background noise of human relations — going at all times, providing an illusion of companionship.

Today's computers, even with their keyboard access and restricted programs, already provide some illusion of companionship (note how words like "delightful" and "cute" are applied to the new round of "user-friendly" machines such as Apple's Lisa, which flashes messages on her screen). Imagine how much more attractive it will be when the computer illusion of human interaction is refined enough to pass for the real thing at its lower levels.

Terminal Loneliness

A common failing of friendship, and a central failing of romance, is that we try to mold others to fit the image we seek, instead of experiencing them as they are. Everyone is guilty of this at some point — of longing for a fantasy, or a convenience, rather than accepting the less perfect but ultimately more rewarding presence of a person's true nature. Molding others in your own image seldom works, however, or at least not for long. People are extremely resistant to such tampering.

But computers aren't. In fact, computers crave it. The computer program is, by nature, the reflected image of the programmer's desires. Powerful small computers can be programmed to behave in whatever way pleases their owners, and as computer power expands, and computers learn to speak, it will be possible to give computers "personality traits", or at least the illusion of them. Again, this has nothing to do with artificial intelligence, merely with chip capacity. A computer could be programmed to show interest in some subjects but not others; to speak in the desired tones of voice; to discuss certain things at certain times of day; to deliver flattery. The computer would only be acting out stage directions, of course, but acting them in a very convincing and convenient way.

In nearly all movie and television science fiction, it's been noted, the computers and the robots have far more personality than the living characters. C3PO was more intriguing and less predictable than Luke Skywalker; the evil robot Lucifer, with his droll wit, had all the good lines in "Galactica"; the kindly robot Robbie nearly stole the show in *Forbidden Planet*. Some of this traces, of course, to the quality of scriptwriting and to reliance on characters so wooden that mechanical men look real by comparison. But some is due to the ways in which computers can be seen as more desirable than people. Computers can offer the appearance of life — the talking, the joking, the interest in what's up — without any of the emotional or spiritual complications. Computers have no feelings, so they can be insulted or mistreated without guilt. They have no rights, so they can be switched off should they grow tiresome. Most important, since they are not living and have no expectations, we incur no obligations to them.

These are more than fine points of the conventions of sci-fi scriptwriting: they reflect on how, today, we are beginning to relate to our electronic companions. And they suggest why many of us might come to view computers as companions of choice. This has already happened in fiction. In Arthur Clarke's new best seller, *2010: Odyssey Two*, sequel to *2001: A Space Odyssey*, Dave, the metamorphized star-baby-astronaut, returns in ghost-like form. Meanwhile HAL, the mad computer, has been cured of his antisocial tendencies by a Freudian psychiatrist. At the end of *2010*, Dave departs for some other-worldly higher plane to await some unspecified great event that will happen near the end of time; to while away the intervening hours, he selects as his companion, you guessed it, HAL the computer.

When the video game industry started, its first products were conceived as moving versions of the traditional game arrangement — two people would play against each other, only through the medium of a machine instead of on a board. Pong, the first video game, matched two or four players, as did almost all of Atari's early games, like Combat (two tanks) and Indy 500 (two race cars). Video game sales didn't take off until the group-play orientation was eliminated and games were designed as man-versus-machine. Space Invaders, Asteroids, Pac-Man, Defender, Pitfall, Galaxian, Frogger — all those that have really made money — are played alone. Computer game manufacturers finally realized that their market appeal lay not in colors and motion, but in the fact that their computers could substitute for an opponent. Conventional games like Monopoly and Risk simply can't be played by a single person and the only play-alone conventional game, Solitaire, has become a cultural synonym for desperation. Computer games circumvent this problem by providing both the means to play alone and the illusion of someone to play with — by bleeping, blinking, changing, and making moves that at least appear spontaneous.

It's all in the scene at a video-game arcade. Dozens of players (mostly teenagers, who should be out petting or playing softball) hunched over joysticks, fully absorbed in private action. No one talks to anyone else — there's so much bleeping and blasting going on, talking is impossible anyway — and no one pays any attention to what others are doing. The prevailing air is anonymity, not too different from the Times Square peep show where customers slink in, slink out, and try not to be noticed while taking anonymous pleasures.

The next wave of video games will advance this condition from teenagers and college kids to adults. Powerful new game-players like the Atari 5200 make possible cartridges that aren't just spin-and-shoot, but that involve strategy, contemplation, and changing circumstances. The first such adult-oriented video game, Atari's Star Raiders, is already on the market, with other slower, more subtle games on the way. Once video games are complex enough to engage the adult mind, computers will be all the more attractive as an alternative to companions.

Advertising the solitary aspects of computers presents a delicate problem, as does advertising one-portion frozen meals and other products that serve the lonely market. A direct approach obviously cannot be used ("NOW! SOFTWARE FOR LOSERS!"). But the way computer game advertisers have already solved the problem indicates how computer-companion advertisers will solve it in the near future. In contrast, say, to beer commercials, like "Welcome to Miller Time", which shows hale friends in hearty communion, video game advertisements rely on grim man-versus-machine confrontations. A commercial for Atari's 5200 has a single young man, alone, matched against the machine in a featureless enclosure. Mattel Intellivision's advertisements offer George Plimpton declaring, in his Duchy of Grand Fenwick accent, that his games are intellectually superior. The most revealing ad, however, was a Plimpton-Mattel commercial run this Christmas. A photo trick made it appear, à la the old Patty Duke show, that there were three George Plimptons in the same room engaged in earnest conversation with themselves about the virtues of various game cartridges. The context was entirely self-centered, as if, through the computer, you could enjoy the company of a group consisting solely of you.

A preview of that culture already exists, in the group now derided as "computer nerds". Computer nerds are the kids who looked funny and didn't fit in during high school, who lost themselves in calculus and physics as an escape, and who came into their glory with the computer revolution. The most celebrated computer nerd, Steve Wozniak, created the Apple I; others have fostered important advances in small computers and video games. While expensive "mainframe" computers and long-term research tend to be in the hands of large corporate teams, it is the nerds, often working alone, who specialize in bringing applied computer technology to the consumer market. The nerds, in other words, are the ones shaping the computers available to the average person, and they are shaping them in their own image — the image of people for whom the personal side of life has often been harsh, and to whom computers, as companions, are a way out.

Yet non-nerds, regular folks who would consider themselves emotionally well-adjusted, comprise the bulk of the computer-buying market. The reason relates, in part, to our educational system. In general, as educational opportunities have expanded, placing college and even graduate school within reach of the many rather than the few, increasing numbers of people have acquired a preference for making life as much like school as possible. It's only natural. If you spend the first 25 years of life in school, being constantly reminded by generous, self-sacrificing parents that school is the only hope of a better tomorrow, when you get to tomorrow you'd like it to be as familiar as possible, and what's familiar is school.

This reflects the tremendous urge of people in my generation to become lawyers, not just because practicing law is financially attractive, but because it's very much like school — writing term papers (briefs) based on library work, being quizzed by professors (judges) on recall of obscure facts, and pretending to care about convoluted theories. It reflects on the present enthusiasm for an "information industry", because a society based on compiling statistics and analyzing reports would make all life like school. And it reflects on the appeal of computers. Computers are a realm in which the mind prevails, where the intellectual facilities my generation has cultivated at such effort and expense can actually be employed. In computers the skills of school — how to study and think — are the skills of success. It's a relief, and more than a little flattering, to finally find an area of life that reveres the brains we always thought we had.

Meanwhile, the pure-logic side of computers appeals to every person who has developed the mind at the expense of the heart. In computers logic is absolute, unchalleangeable; no chance circumstances or inconvenient emotional factors intrude. It is often observed that intellectual prodigies are slow to mature emotionally, because they dedicate themselves to abstract pursuits and are angered, rather than intrigued, by the concrete world and its compromises. With the great college expansion having taught so many to love the abstract and scoff at the real, even those who don't have unusual IQs or artistic gifts may slide into arrested emotional development. Almost anyone who has done well at any level of the now-large educational system, and been lauded by parents or teachers for doing so, is in great danger of mistaking the mental world for the real world. To them, computers have become a necessity — a way to show that academic skills *are* the real world, by going to a place where only brainpower matters.

This sheds light on part of the business community's hunger for computers. Computers obviously are valuable to many companies, and as time passes and computers improve, will become more so. But it's a safe bet that many computers are being bought for businesses that don't really need them or can't justify the cost, and that management-school graduates are to blame. Modern management schools are not all that different from philosophy departments; students are taught to think in terms of models and logic, to admire detached analysis and brainpower more than experience and to toss out any earthly complications that interrupt the flow of theory. When management-school graduates move into the business world, they soon discover a bewildering array of emotional, unpredictable, and just plain illogical forces that their textbooks said they could assume away for the sake of a flawless model. A sales presentation may be quite well-done, for instance, and still not work, even though it "should". So modern managers are more and more willing to retreat to the realms where "should" is the only factor — computers. By computerizing everything in reach the modern business manager can see {his} logical faculties rewarded and can put off confrontation with the intractable, illogical problems that may be what really need {his} attention. Similar events are taking place in government, the military, and other fields of life.

Perhaps oddly, even the dropouts of the sixties are making their peace with computers. Peter McWilliams, author of the hot-selling *The Personal Computer Book*, got his start by writing about transcendental meditation. David Sudnow, author of *Pilgrim in the Microworld*, was a sociology teacher at Berkeley during the free-speech days. Computer firms, Susan Chace reported recently in *The Wall Street Journal*, are enjoying substantial sales to counter-culture types. Computers, perhaps, have some appeal here as the first "clean" industrial technology (users don't have to see the smokestacks of the generating stations that provide the computers' power) and because of their ability to expand the mind in a way reminiscent, minus the physical peril, of drugs. The computer is an ethereal world that welcomes creativity — not exactly what Timothy Leary had in mind, but better than nothing. My speculation, however, is that the hippie heart responds to computers mainly in emotional ways. The counter-culture was rejected and misunderstood by the mainstream. With no Woodstock to retreat to, computers that can be programmed to give at least the illusion of understanding will have to do.

The Fido 5200

For the lonely and the overly intellectual of this generation and others to follow, computers could be the main agents of comfort and consolation. All things considered this is not half bad. Computers are safer than drugs or drink; they have worthwhile applications; they're smarter than dogs or parakeets. Yet there is something they can never be, regardless of the level of technological advance. A dog's emotions, though highly simplified compared to ours, are nonetheless *real*. A dog really *cares*; when you're away he misses you and when you're present he feels joy. These emotions are amplified by numbers. The more people who are around, the happier the dog is. People are the same, happiest in the company of family and friends, although it's sometimes harder to detect since people don't display their emotions in the exaggerated style practiced by dogs.

Obviously, on this scale, the most expensive computer finishes way behind a labrador. A computer's reactions aren't real, merely realistic. Maybe for a lost soul, badly hurt by the world and its unreliable occupants, computers are a haven we should not begrudge. But what happens when people start going straight to computers for their companionship, bypassing the danger of hurt but also the hope of transport?

Technostress

CRAIG BROD

The difference in the ways children lose themselves playing baseball and using computers, is striking. Baseball, like other typical children's pursuits, is stimulating in a variety of ways. Baseball players are interacting with each other, learning about teamwork and competition, enjoying exercise, chatter, and camaraderie, honing eye-hand coordination and peripheral vision, coping with quiet spells and pressure situations, learning to accept defeat and victory gracefully. It is, in sum, a social activity. Computer work, by contrast, is usually a solitary, antisocial pursuit, generally devoid of demands on the imagination. Outside stimuli are shut off. Because parents think that refining computer skills is advantageous for their children, they often approve when their children are lost in the computer time warp; but they ignore the cost to the child.

Piaget was fascinated by the way children view time, and he devised an experiment to study the question. He would show a child two sequences of pictures in rapid succession, first sixteen pictures in four seconds and then thirty-two pictures in four seconds. He found that children younger than eight years old tended to think the second, more rapid sequence took longer. Children eight and older generally perceived the reverse: more events within a time span seemed to take less time.

The essence of the computer is speed. Because events occur with great frequency, their duration appears shorter. There is only a stream of mental events, without the dilution of motor activity. The child must constantly make decisions, choose directions, react to outputs on the screen. The computer user cannot stray too far from the prescribed logic of orderly, well-structured procedures, or the computer will not respond. The speed and intensity of this activity heightens the sense of engagement the child experiences with the work. In general, there is a reduction of external sensory experience. The outside world fades, and the child becomes locked into the machine's world.

Children cannot reduce their sensory inputs without closing off their interactions with others, and, as with adult computer workers, time distortion and social isolation go hand in hand. Parents may notice that a problem exists when a child has difficulty shifting contexts, from computer to people, or from computer to homework. The computer involvement begins to color all other interactions. In extreme cases, children develop an intolerance for human relationships. They become accustomed to high-frequency logic, a rapid-fire dialogue between screen and fingertips that makes time seem to speed up. In talking with parents, siblings, even friends, time drags by comparison. The only way to make the transition from machine to human is to talk about computers.

Ian (age 13): I feel teachers talk too much. They could say half as much and be more efficient. The one I had last year just liked to lecture and never got to the point. In regular conversations you're just doing something to use up time. You don't do it unless you've got time to waste.

James (age 16): Sometimes it's hard to switch over [from programming to family activities]. Work with the computer is like being in a bubble. Once my bubble's broken all the liquid flops out, and then I can be outside again. I shake once or twice, and I'm back in the real world again, trying to function like normal.

Matt (age 16): Everything I say isn't . . . computer this and computer that . . . . but a lot of it is because my friends know what I'm talking about . . . . This gives me satisfaction. I don't have to talk things out a lot. It's like writing things out longhand — I don't like doing that either. I like to do things faster and better.

For young, intensely involved computer users, this altered sense of time is also changing their attitudes toward traditional learning media, such as books.

Tom (age 13): You know, a computer is more like real life. Real life is something that's actually happening. In a way, books are real life because you're thinking about them happening while you're reading, but in a computer, you're actually doing it instead of reading about something that's happening. You're there in a computer. You're part of what's happening, and it's faster . . .

In making the shift from the computer to other contexts, one is forced to come to grips with the intangibles of the world — subtlety, reflection, sensory awareness, imagination. Children perceive this requirement as slowing them down. When a child is interacting with a computer, everything "clicks". Barriers between the child and the activity are at a minimum. The computer seems to pull the child right in.

In *On the Experience of Time*, Robert Ornstein notes that we normally use coding schemes to save energy when we digest information from the world. For example, if we hear the sequence of numbers 149217761945 read to us and are asked to remember them, we will have difficulty — unless we are told they are the dates Columbus discovered America, the Declaration of Independence was signed, and World War II ended. It is easier to remember three chunks (and familiar ones) than twelve random ones. Coding tricks like "chunking" apply to higher-level information and events as well. Human communication is complex. The codes to understand it are not always clear to adults, and they are less so to children. Human communication is difficult to "chunk" — it is inefficient, subject to interpretation, and hard to categorize. Since children don't have the experience or background to use human coding, they adapt more quickly to the sparse and uncomplicated coding of the computer. In conversation with an adult who does not speak in short, clear bits, they are more apt to be frustrated.

The same is true of reading. Reflecting on a story, imagining what the characters are going through, and wondering about the story's meaning all require the brain to process a great deal of information, and this can make time seem to slow down. For some children, this is a disincentive. The outside world provides challenges that require creative energy and new resources; the computer provides efficient communication. Kids being raised by machines prefer efficient communication.

The Technological Threat to Education

ROBERT J. SARDELLO

I am going to talk about computers and the promised revolution in education attendant on the arrival of a promised computer culture. I want to expose the utopian fantasies inherent in all talk of computers' revolutionizing education, and right at the start, so you will know where I stand. I am firmly opposed to the introduction of the computer as a technological device oriented toward changing the very tradition of education. Now, this may seem to be a very close-minded position, not at all like a practicing psychologist who almost always has trouble saying whether anything is right or wrong. But this evaluation is not a prejudice, a prejudgement. It comes out of a great deal of consideration of what kind of thing the computer is, and a great deal of consideration of the relation between technology and culture. And the conclusion that I have come to may surprise you. The computer, if it is allowed to infiltrate the very heart of education in the particular manner I will outline in a moment, will destroy education: not because it is a mechanism, and as such threatens to transform human beings into likenesses of itself; the destructive power of the computer is to be found in the fact that it transforms education into psychology. Furthermore, the peculiar kind of psychology that characterizes the computer is a psychological symptom, and the symptom is that of the psychopath. In other words, if the kind of computer education that I am about to outline actually comes about, we will be transformed into a culture of psychopaths.

So, having started with a conclusion, a conclusion that I trust has evoked your attention, let me try to demonstrate that I know whereof I speak. First, let us establish the particular kind of computer education that is a threat to the whole of Western culture. Computers used as technical devices to perform operations that are themselves technical are not included as threatening to culture. Pocket calculators, word processors, and all the variety of programs that are ready-made for use in personal computers do not pose a threat to the very meaning of education. These technical devices can free the imagination for the consideration of matters involving mathematics, accounting, economics, or business. Or, in the case of word processing, the imagination is set free to focus on the craft of writing itself.

What is called "computer-assisted instruction" also does not pose a real threat to culture. Teaching machines or programmed instruction have already shown themselves to be dismal failures, precisely because they turn the learner into a mechanism, who duly responds with frustration and boredom.

The technological threat to education is to be found in the claim that teaching the child to program the computer can be done in such a manner that programming teaches the processes of thinking itself and thus removes the necessity of formal classroom instruction. Such a claim has been put forth by Seymour Papert, the Cecile Greene Professor of Education at M.I.T., inventor of the system of programming called LOGO, which is rapidly finding its way into educational settings. These are the claims set forth by Papert:

I see the classroom as an artificial and inefficient learning environment that society has been forced to invent because its informal environments fail in certain essential domains, such as writing or grammar or school maths. I believe that the computer presence will enable us to so modify the learning environment outside the classroom that much if not all the knowledge schools presently try to teach with such pain and expense and such limited success will be learned, as the child learns to talk, painlessly, successfully, and without organized instruction. This obviously implies that schools as we know them today will have no place in the future.[1]

Of course, LOGO can help in the teaching of traditional curriculum, but I have thought of it as a vehicle for Piagetian learning [that is, following the system of the Swiss psychologist, Jean Piaget], which to me is learning without curriculum.

I see Piaget as the theorist of learning without curriculum and the theorist of the kind of learning that happens without deliberate teaching. To turn him into the theorist of a new curriculum is to stand him on his head.[2]

But "teaching without curriculum" does not mean spontaneous, free-form classroom or simply "leaving the child alone". It means supporting children as they build their own intellectual structures with materials drawn from the surrounding culture.[3]

These claims are radical. We need to be prepared to accept them if they can indeed produce a genuine reform in public education. The radicality of the claims makes them most appealing. There is no patchwork on a system of education dictated by the system itself here, with all the built-in protections to assure that the system remains relatively intact. These claims offer a true revolution in education. Is it, however, a revolution that renews the roots of culture and tradition or one that will destroy the memory of culture altogether?

Teaching children to program computers in school does not have as its aim the introduction of a new subject matter into education. Computer programming, as outlined in the LOGO system, is not subject-centered but child-centered. It is the ultimate extension of the methods course now so prevalent in teacher education, for, in a significant way, it eliminates content altogether and reduces all education to method; all things can be learned through the method of computer programming. So, it is necessary to be aware of the false claim that argues that computers must be introduced into schools because we are entering a time when anyone who does not know programming will be illiterate in a computer culture. The very term *computer literacy* establishes the computer as far more than a machine; such a term is an assault on the basis of culture itself by assuming that computer programming is truly a language — not a constructed, artificial language, but a language fully capable of expressing the full and complete life of a people. There is a huge inflation to the suggestion that the artificial terminology of computer programming constitutes a language at all. It assumes that language no longer emanates from the life of a community, from geographical place, from heritage, ritual, from the living body of a people, from the voice of things in the world. The new origins of culture, so the inflated claim would state, are IBM, Apple, Xerox, and Texas Instruments. Computer terminology is certainly not a living language, but rather the enslaving of language by turning every form of speech into an object to be manipulated by the totalitarian grammar of computational logic.

The computer looms before us as far more than a device for presenting, in "programmed style," traditional subject matter to be learned. The claim that we are entering into a computer culture must be clearly understood and taken quite seriously, in spite of all its falseness. Culture is never a progressive affair. Culture always comes about by looking backward, by recovering the past, by relating the present to permanent patterns of the soul, by remembering the dead, by reflection of values. The aim of introducing programming into the schools is to forget the past, not to renew it.

We must realize, however, that the claims of programming carry any persuasive power at all, because true, living culture is itself nearly dead. Education has not done its job, the primary job of schooling, which is the initiation of the young into the life of culture. As Donald Cowan pointed out in his lecture on the economics of taste last summer, the techno-economic sphere of society has already invaded the cultural enterprise of education. Education models itself on the pattern of corporate mentality. Administrators are not the intellectual leaders of the school, but the managers of a system. And teachers are not considered to have the inner authority of those who follow a calling in life, a vocation, but are considered to be staff. It is a short hop, skip, and jump from that situation into the replacement of staff by computers. The technological conception of education took over the schools quite some time ago. Moving in the devices cannot be seen as an innovation; rather, it brings a technical vision to completion. Computers would not have a place in schools had schools not already become possessed by the technical imagination.

How will computer programming eliminate curriculum in the schools? Programming as a method of learning has entered the school through the introduction of what Papert calls "the LOGO environment". The aim of this computer environment is to replace curriculum. For example, currently, in schools, one of the subjects studied is English grammar; it is part of the curriculum. But grammar is not a subject that interests students very much; for most teachers, it is probably a pretty deadly subject to teach. The life has gone out of the subject. Because teachers themselves do not find life in such subject, students find it totally unrelated to life. If teachers sensed the life of grammar they would certainly organize and demonstrate against the threat of having something they love removed from them. Teacher education courses have devised methods courses for the teaching of grammar, but with the concentration on method — in attempting to make the subject lively and interesting for the student, more and more life is drained from the thing itself. It is an instance of the now universal occurrence in which teachers themselves do not learn grammar and in so doing find the life within the things, but instead learn all kinds of techniques of infantilizing grammar to make it suitable to the psychology of the child. The attempt is to care for the child more than the thing, which is really a terrible kind of psychology because real learning is possible only by becoming the thing, not by turning the thing into a subjective child-centered psychological process.

Grammar has its own kind of life; it has a soul of its own. It has invisible spirits within it. Nouns, verbs, adjectives, adverbs, sentences, are extraordinary beings, full of character, personality, color, depth, imagination. And they become even more interesting when they get together and shape a world, when they speak in unison, full of tension, vibrancy, resonance, each giving up a little of itself to make another word show hidden parts of itself. Of course, it takes a good deal of rote memorizing to learn all the parts of speech and a good deal of diagramming to learn to see the dance of sentences. Such memorizing used to be called learning by heart; it is learning through the heart. One learns by giving one's heart to the thing. We need a full psychology of grammar. It is tragic indeed that the focus is on the psychology of the child learning grammar. That psychology kills the thing. Grammar, of course, mightily resists such defilement, and in spite of all of the efforts of the methods courses, in fact perhaps because of them, grammar becomes increasingly difficult to learn, requiring a repetitious, detached, seemingly meaningless immersion in definitions, rules and exercises, which can then be applied to actual writing. That is to say, the more concentration on method, the further the life of grammar recedes. The introduction of the computer takes this direction to its final conclusion by eliminating grammar altogether.

The LOGO approach to the teaching of grammar through computer programming is that grammar is not studied at all. Grammar is eliminated from the curriculum. The subject would be replaced by the programming of a general structure, for example, the structure of a poem, within which words are selected from a random list to fill in that structure. The student plays around inserting various words into the general structure until an error-free poem is produced — that is, one in which each part of grammar is correct. The aim is to bypass grammar because grammar is not interesting; but a child can feel like a poet by writing a poem, and, incidentally, effortlessly learns the parts of speech and what they do while constructing a poem.

The argument is in fact quite persuasive. When we learned to talk in our early years, none of us did so by studying vocabulary and the parts of speech. We would say something and perhaps be told that we had used an incorrect word or a word in the wrong place. Gradually, we learned to speak more or less correctly. Now, what is the difference between that kind of learning and the learning proposed here? Most obviously, learning how to talk occurs in the context of a family, a community, through the mediation of the breath and the heart of those with whom we are most intimately associated. Remove any such community or put them aside from the actual act of learning and into the role of cheerleaders and what do you have? I can best show the result by considering a poem generated by a thirteen-year-old girl, or, should we say, generated by a computer. This poem, reported and discussed by Seymour Papert, was produced by inserting words into a programmed structure one by one. The program would accept some words and reject others until the structure was filled out. If, for example, a noun was selected where the structure called for a verb, there would be some error message flashed on the screen and the student would try again. The error message, incidentally, would not be a judgement — in computer language there are no rights and wrongs, because, my heavens, we would not want the child to think badly about {his* performance! When an error message comes up — something like "that won't work, will it? Let's try again. I can't fit that word in here" — it is necessary to "debug" the program. At any rate, here is the thing produced by the child:

Insane Retard Makes Because Sweet Snoopy Screams

Sexy wolf lives thats why the sexy lady hates
Ugly man loves because ugly dog hates
Mad wolf hates because insane wolf skips
Sexy retard screams thats why the sexy retard
Hates
Thin snoopy runs because fat wolf hops
Sweet foginy skips a fat lady runs[4]

First, of course, this thing is not in any sense a poem, not even a failed poem. It is not that there is any lack of rhyme, rhythm, meter. None of that is really necessary for a poem. It is rather that this thing does not compel any attention. It has no life of its own. It has only the attention that anyone is willing to give it, and I have already given it far too much. Second, the words of the poem do not speak to each other to make anything that speaks. There is no internal tension between the words. It is a thing that imitates a poem but does not have the heart of a poem. However, the educationalists of programming might well agree that this is not a poem. The point, it would be argued, is that the child thinks that she is writing a poem; that makes her feel important, like an adult who is

*See "Questioning Technology, Questioning Patriarchy," p. 1.

actually doing something meaningful. Actually, while she thinks she is writing poetry she is effortlessly, without thought, learning the parts of speech. Making something that looks like a poem is just the excuse for secretly teaching grammar without grammar curriculum.

It is just here that the psychopathic character of programming as a method of learning begins to show its disguised ugliness, how education has gone from the order of the heart to the order of calculative manipulation. In order to highlight this psychopathy, it is necessary for a moment to describe this psychological illness in some detail. I am neither leaving the subject of computer learning nor imposing a psychological category onto programming, but rather describing the inherent psychology within programming as a method of learning.

Psychopathy is actually a kind of programming in life, learning how to debug life. The psychopath does everything effortlessly, freely, without any sense of inhibition, restraint or suppression. Nothing of the world makes a claim on the soul of the psychopath. Cheating, lying, saying one thing and doing just the opposite without least concern, changing a position from one moment to the next in order to satisfy the situation, the psychopath is always a winner. In appearing better than one actually is, successfully gliding along the surface, intelligently but without insight, without being emotionally moved, without feelings of the heart, only programmed feelings to suit the situation, psychopathy constantly assures that everything works smoothly, efficiently, always to one's advantage. Everything is a game — feelings, emotions, courtesy, love, sympathy for others, expressions of care. Like the little girl's poem, which looks like a poem but does not have the heart of a poem, the psychopath can imitate any form of behaviour without it's going through the heart.

Look at the word *psychopath* — the pathos of soul. It is different from the word psychotherapy, which is the logos of the suffering of the soul. Why is this illness called psychopathy? Because it describes the situation of the soul that is unaffected by anything. The psychopath lacks any erotic connections with the world. The illness is one of constant manipulation of one's own psychic processes for the satisfaction of the moment. The center of gravity is totally on the side of oneself. Is this not like the child who looks at the computer screen and sees there not a content, a curriculum that one must become in order to learn, but rather sees only his own psychological processes as they are happening each moment, and is given the task of manipulating them? Of course, it is fascinating, absolutely captivating to stare at oneself displayed on a screen for hours on end. Of course, it is intriguing to control one's own processes, making them do whatever is necessary to receive a result. Of course, this is exciting. But is has no heart.

This game of computer programming has nothing to do with content, with something that has a soul utterly different from our own in which learning involves learning to love the soul of that which we do not understand. The word *education* means a leading out of the soul; and that says learning consists of allowing our soul to enter, via the heart, into the things of the world, to mingle there in unfamiliar conversations, to be affected, moved by the soul of things, gradually learning to speak their language.

The roots or psychotherapy are found in the loss of care for the things of the world. Learning by the method of programming concentrates on three components — the ego, the body, and an abstract sign on the computer screen. Seymour Papert names this sign a "turtle", which is no more than a kind of arrow on the screen that traces out geometric shapes according to computer commands. The programmer decides what he wants to construct, an operation of the ego, and relies on a vague memory of the body to determine how to move the arrow on the screen forward, backward, right, left, up and down. The object of computer learning is to remove the child from the actual world and to insert {him} into {his} own subjective processes where an imitation world is invented. A recent "Nova" television program devoted to the work of Seymour Papert provides a series of striking images of children learning to turn away from the world.

A first scene. A group of children stand out in an open field. The field is beautiful — tall green grass, purple and yellow flowers beneath a deep blue sky. The children are not playing in this field: in fact, they are quite oblivious to their wonderful surroundings. It is quite an extraordinary image. Imagine standing outside on a cool autumn morning; there is dew on the grass; the clouds play coy games, making themselves into shapes of monsters, old men with flowing white beards, beautiful princesses. The air, pure and cool, draws deep breaths. The coolness says that in a short time this grass will go to sleep for the winter; this may be the last time for months to romp and play, turn somersaults. Even an adult could not resist being moved to play, or at the very least to walk and contemplate the change of seasons. But these children, ι m ιved by such beautiful things, walk off geometric shapes through the grass. One little girl is the turtle; her directions — ten steps forward, turn left, then another ten steps straight ahead, another left turn, ten more steps, and a final sharp left turn. The girl is blind to the world. And so are the rest of the children, who are all calculating the movements necessary to make a square. When they have completed the motions, they all go inside and sit in front of the computer and program the turtle on the screen to trace out the shape of a square. This scene is paradigmatic of computer learning. It is all an abstraction. learning does not celebrate the actual things of the world, but turns away from the world in order to program an imitation world.

Another scene. Two small girls dressed in black leotards sit cross-legged on a stage, facing each other. Music starts. The girls spin around on their bottoms several times, rise from the floor, twirl several times past each other and then cartwheel back, crossing each other again. It is all perfectly executed. It looks like a dance. It is a perfect imitation of a dance. Yet, it is not a dance. Absolutely no tension shows on the faces of the girls, and there is no tension in their bodies, and no tension between their bodies, in the space between their movements. It is a perfect series of movements, but it is not interesting to watch. The viewer is not moved by the dance. There is no edge to it, no fearful, exciting feeling that the girls are right on the line between an earthly form, the natural movements of the body, and the transformation of the body into the world of dance. The girls had programmed the entire sequence of movements on the computer, analyzing each movement necessary, and then followed this program to execute the motions. It is a new ballet — the dance of the psychopaths.

A third scene. A group of children are standing outside a building, evidently their school. They are crowded together watching a juggler. First he juggles four balls — red, green, yellow, blue. Then he juggles four pins. They fly through the air, not a one at rest. It is marvellous, magical. The children, however, are not delighted. How is it that they are not squealing, spellbound with such a wonder? The camera focuses on one little boy. His face is contorted with concentration. Eyes sharply focused, his head rapidly moves, following the motions of the balls and pins, carefully trying to dissect each movement of the juggler. The carnival is missed. The boy is already calculating the necessary commands to program the motions into the computer. Once done, he will program his body to carry out these same analyzed movements. He will be a perfect juggler, a psychopathic juggler.

Each of these portraits, along with our example of the girl's programmed poem, seems to be an instance of learning, each a rather remarkable accomplishment. And even if they lack that central dimension of heart, even if children are numbed, anesthetized to the world, are not the accomplishments the true measure of learning? Are not these children being prepared to enter the world fully capable of mastering any task quickly, efficiently, and perfectly? Is not the absence of heart that is so apparent in the actions of these children nothing more than a sentimental attachment to a world that no longer exists anyway? I can only answer by saying yes, indeed, this method of learning suits the culture we have constructed. The age of psychotherapy is in its prime. But there are two reasons why this world, dominated by the technical imagination, is unacceptable, and must be reversed. Lurking within every psychopath is suicide, depression, and violence. Here is the price that we shall have to pay for quickly won perfection lacking the beauty and the heart. A massive bomb is in the making, and while we have made the bomb, it is our children who will explode it at the very moment when life, as it always does, gets its way, and manipulative calculation does not work. We see it already brewing in our children. The moment their calculations in life do not achieve the desired result, they are quick to violent anger.

The second reason is far more important. If, for a moment, it is possible to realize that education is really not for the sake of persons at all, but rather for the sake of the world, it suddenly becomes clear that child-centered learning is a preparation for the destruction of the world. Of course, we are filled with fears of the holocaust; the holocaust is already upon us. We shall be all dead before the bomb goes off.

I end with a final image. The film *Bladerunner* portrays the city of Los Angeles in the not-too-distant future. The image of that city ought to strike horror in our hearts. The city is in ruins. So much pollution fills the air that it is constantly raining. Most of the buildings are abandoned, as if some surge of overdevelopment by speculators did not bear fruit. Abandoned automobiles fill the streets. Garbage is scattered everywhere. Seamy characters roam the city. Everything is in utter ruin. Layered on top of this image lies a world of technological perfection. Automatic, air-lock doors open and close along the street. A huge dirigible passes over the city periodically, neon sign advertising some product. Police patrol the ruins in electronic airships. Lights, electronic gadgets, computers of every sort populate this city, a massive image of turning our back to the world in order to construct an imitation world. The main character of the film has the task of searching out a group of genetically engineered creatures who look exactly like human beings except that they are perfect — they are brilliant, have incredible athletic bodies, are extremely violent, seeing no reason for refraining from killing anyone who blocks their psychopathic wishes.

My thought is simply this: How could it ever be that the world could matter so little that it becomes a garbage heap in the midst of incredible technical achievement? And, as I look at the beautiful, perfect, emotionless, heartless creatures, I fear that there is no need to develop genetic engineering to produce them. They may well be the product of public education.

Notes

1 Seymour Papert, *Mindstorms: Children, Computers and Powerful Ideas* (New York: Basic Books, 1980), p. 8-9.
2 Ibid., p. 31.
3 Ibid, pp. 31-32.
4 Ibid.,p. 49.

{Man}Bytes Dog[*]

JAMES GORMAN

Many people have asked me about the Cairn Terrier. How about memory, they want to know. Is it IBM-compatible? Why didn't I get the IBM itself, or a Kaypro, Compaq, or Macintosh? I think the best way to answer these questions is to look at the Macintosh and the Cairn head on. I almost did buy the Macintosh. It has terrific graphics, good word-processing capabilities, and the mouse. But in the end I decided on the Cairn, and I think I made the right decision.

Let's start out with the basics:

MACINTOSH:
Weight (without printer): 20lbs
Memory (RAM): 128 K
Price (with printer): $3,090

CAIRN TERRIER:
Weight (without printer): 14lbs
Memory (RAM): Some
Price (without printer): $250

Just on the basis of price and weight, the choice is obvious. Another plus is that the Cairn Terrier comes in one unit. No printer is necessary, or useful. And — this was a big attraction to me — there is no user's manual.

Here are some of the other qualities I found put the Cairn way out ahead of the Macintosh:

PORTABILITY: To give you a better idea of size, Toto in "The Wizard of Oz" was a Cairn Terrier. So you can see that if the young Judy Garland wss able to carry Toto around in that little picnic basket, you will have no trouble at all moving your Cairn from place to place. For short trips it will move under its own power. The Macintosh will not.

RELIABILITY: In five to ten years, I am sure, the Macintosh will be superseded by a new model, like the Delicious or the Granny

[*] See "Questioning Technology, Questioning Patriarchy," p.1.

Smith. The Cairn Terrier, on the other hand, has held its share of the market with only minor modifications for hundreds of years. In the short term, Cairns seldom need servicing, apart from shots and the odd worming, and most function without interruption during electrical storms.

COMPATIBILITY: Cairn Terriers get along with everyone. And for communications with any other dog, of any breed, within a radius of three miles, no additional hardware is necessary. All dogs share a common operating system.

SOFTWARE: The Cairn will run three standard programs, SIT, COME and NO, and whatever else you create. It is true that, being microcanine, the Cairn is limited here, but it does load the programs instantaneously. No disk drives. No tapes.

Admittedly, these are peripheral advantages. The real comparison has to be on the basis of capabilities. What can the Macintosh and the Cairn do? Let's start on the Macintosh's turf — income-tax preparation, recipe storage, graphics, and astrophysics problems:

	Taxes	Recipes	Graphics	Astrophysics
Macintosh	yes	yes	yes	yes
Cairn	no	no	no	no

At first glance it looks bad for the Cairn. But it's important to look beneath the surface with this kind of chart. If you yourself are leaning toward the Macintosh, ask yourself these questions: Do you want to do your own income taxes? Do you want to type all your recipes into a computer? In your graph, what would you put on the x axis? The y axis? Do you have any astrophysics problems you want solved?

Then consider the Cairn's specialities: playing fetch and tug-of-war, licking your face, and chasing foxes out of rock cairns (eponymously). Note that no software is necessary. All these functions are part of the operating system:

	Fetch	Tug-of-war	Face	Foxes
Cairn	yes	yes	yes	yes
Macintosh	no	no	no	no

Another point to keep in mind is that computers, even the Macintosh, only do what you tell them to do. Cairns perform their functions all on their own. Here are some of the additional capabilities that I discovered once I got the Cairn home and housebroken:

WORD PROCESSING: Remarkably the Cairn seems to understand every word I say. He has a nice way of pricking up his ears at words like "out" or "ball". He also has highly tuned voice-recognition.

EDUCATION: The Cairn provides children with hands-on experience at an early age, contributing to social interaction, crawling ability, and language skills. At age one, my daughter could say "Sit", "Come" and "No".

CLEANING: This function was a pleasant surprise. But of course cleaning up around the cave is one of the reasons dogs were developed in the first place. Users with young (below age two) children will still find this function useful. The Cairn Terrier cleans the floor, spoons, bib and baby, and has an unerring ability to distinguish strained peas from ears, nose and fingers.

PSYCHOTHERAPY: Here the Cairn really shines. And remember, therapy is something that computers have tried. There is a program that makes the computer ask you questions when you tell it your problems. You say, "I'm afraid of foxes". The computer says, "You're afraid of foxes?"

The Cairn won't give you that kind of echo. Like Freudian analysts, Cairns are mercifully silent; unlike Freudians, they are infinitely sympathetic. I've found that the Cairn will share, in a nonjudgemental fashion, disappointments, joys and frustrations. And you don't have to know BASIC.

This last capability is related to the Cairn's strongest point, which was the final deciding factor in my decision against the Macintosh — user-friendliness. On this criterion, there is simply no comparison. The Cairn Terrier is the essence of user-friendliness. It has fur, it doesn't flicker when you look at it, and it wags its tail.

III

TECHNOLOGY
The Web of Life?

8. *Contemporary society can be described in phrases like "Information Age" and "global communications network". What do the "information" and "comunications" consist of?*

The acceleration of technology has been accompanied by an eclipse of meaning, as ideas have passed from understanding to knowledge to data. The buzzword "information" signifies this removal of meaning, and the "communication" comes to be mainly one-way, as we spend more and more time passively plugged in to the programming and circuitry of the new order.

It is not difficult, or unrealistic, to dream up a "Big Brother" intentionality out of all this. In fact, some promoters of the Information Age sound pretty sinister. Consider Pask and Gordon (*Micro Man*, pp. 216-217): "Pockets of a de-informationalized society may survive ... But most communities — particularly large prosperous ones — have no choice in the matter. They must opt in. The sooner this fact and its consequences become part of our consensual reality, the better for everyone."

Is this an offer we can't refuse?

French post structuralist Baudrillard argues that meaning

157

and communication actually evaporate in the face of expanding "information". Langdon Winner debunks a number of serious misconceptions, and Herbert Schiller sees the new information technology as a weapon in the armory of the multinational corporations.

The Implosion of Meaning in the Media and the Implosion of the Social in the Masses

JEAN BAUDRILLARD

We live in a world of proliferating information and shrinking sense.[1]
Let us consider three hypotheses:
1. Either information produces meaning (the negentropic factor), but wages a losing battle against the constant drain of sense which is taking place on all sides. The wastage even outstrips any effort to reinject meaning through the media, whose faltering powers must be bolstered up by appealing to a productivity of the base. We are dealing with the ideology of free speech, of media reduced to innumerable individual broadcasting units — "anti-media" even.
2. Or, information has nothing to do with meaning. It is qualitatively different, another kind of working model which remains exterior to meaning and its circulation, properly speaking. This is Shannon's hypothesis: that a purely instrumental sphere of information exists which implies no absolute meaning and which cannot itself, therefore, be implicated in any value judgement. It is rather a sort of code, similar perhaps to the genetic code, simply itself, while sense is an entirely different thing, an after-effect, as it is for Monod in *Le Hasard et la nécessité*.[2] Were this so, there would be no

---

1 Except for this instance, where, for the sake of euphony, I have translated it as "sense", I have preferred "meaning" as a rendering for the French *sens*. — Trans.
2 Jacques Monod, *Le Hansard et la Nécessité* (Paris: Editions du Seuil, 1970).

significant relationship between the inflation of information and the dwindling of meaning.

3. Or, on the contrary, there is a strict and necessary correlation between the two, to the extent that information destroys or at least neutralizes sense and meaning, the loss of which is directly related to the dissuasive and corrupt action of media-disseminated information.

This last option is the most interesting, but it runs counter to received opinion. Socialization is measured everywhere by the degree of exposure to mediated messages, hence underexposure to the media is believed to make for a de-socialized or virtually a-social individual. Information is everywhere supposed to produce an accelerated circulation of meaning, which appreciates in value as a result, just as capital appreciates as a result of accelerated turnover. Information is presented as being generative of communication, and in spite of the extravagant waste involved, the consensus is that over all a residue of meaning persists, which redistributes itself among the interstices of the social fabric, just as, according to the consensus, material production, in spite of its dysfunctions and irrationality, results, nonetheless, in increased wealth and a more highly developed society. We all pander to this myth, the alpha and omega of our modernity, without which the credibility of our social organization would collapse. The fact is, however, that it is already collapsing, and for this very reason, because whereas we believe that information produces meaning, communication, and the "social", it is exactly the opposite which obtains.

The social is not a clear and unequivocal process. The question arises whether modern societies are the result of a process of progressive socialization, or *de*-socialization? Everything depends on the meaning of the term socialization, whose various meanings are unstable, even reversible.

Thus, for all the institutions which have marked the "progress" of the social (urbanization, centralization, production, work, medicine, education, social security, insurance, including capital itself, doubtless the most powerful medium of socialization), it could be claimed that they at once produce and destroy the social.

If the social is made up of abstract demands which arise, one after the other, on the ruins of the ritual and symbolic edifice of earlier societies, then these institutions produce more and more of them. But at the same time they sanction that wasting abstraction, whose specific target is perhaps, the very marrow of the social. From this point of view one could say that the *social* regresses in direct proportion to the development of its institutions.

This process accelerates and reaches its peak in the case of the mass media and information. *All* the media and *all* information cut both ways: while appearing to augment the social, in reality they neutralize social relations and the social itself, at a profound level.

Information devours its own content; it devours communication and the social, for two reasons:

1. Instead of facilitating communication, it exhausts itself in the *staging* of communication. Instead of producing meaning, it wears itself out staging it. This is the gigantic simulation-process with which we are familiar: non-directive interviews, phone-ins, all-round participation, verbal blackmail — "you're involved, the event is you", etc. The domain of information is increasingly invaded by this kind of phantom content, a homeopathic graft, a fantasy of communication. It is a circular arrangement by which the audience's desire is staged, an anti-theater of communication, which, as we know, is never anything more than the recycling of the traditional institution in negative form, the integrated circuit of the negative. What energy is expended in order to keep this sham at arm's length, to avoid the brutal de-simulation which would bring us face to face with the reality of a radical loss of meaning!

It is futile to wonder whether it is the loss of communication which puts this sham at a premium, or whether the pretense is there from the start, fulfilling a dissuasive purpose: that of short-circuiting in advance any possibility of communication, a precession of the model which eradicates the real. It is futile to wonder what the first term is, there is none — it is a circular process, that of simulation, of the hyperreal. Hyperreality of communication and meaning: by dint of being more real than the real itself, reality is destroyed.

Thus both the social and communication function in a closed circuit like a decoy which assumes the power of a myth. Faith in the existence and value of information is confirmed by that tautological proof that the system provides of itself, by reduplicating, through the medium of signs, a reality that is in fact undiscoverable, chimeric.

But what if this faith is just as ambiguous as that which was inspired by myth in archaic societies? One both believes it and doesn't believe it at the same time, without questioning it seriously, an attitude that may be summed up in the phrase: "Yes, I know, but all the same." There is a kind of reverse shamming among the masses, and in each of us, at the individual level, which corresponds to that travesty of meaning and communication in which we are imprisoned by the system. The tautology of the system is met by ambivalence, dissuasion by disaffection or by a belief that is at least enigmatic. The myth exists, but one must beware of thinking that people believe in it: that is the trap of critical thought, which can only operate on the assumption that the masses are naive and stupid.

2. Behind this exaggerated staging of communication, the mass media, the continuous build-up of information, pursue their relentless destructuring of the *social*.

A bombardment by signs, which the masses are supposed to echo back, an interrogation by converging light/sound/ultra-sound waves, linguistic or light stimuli exactly like distant stars, or the nuclei that are bombarded by particles in a cyclotron: that is information. Not a mode of communication or of meaning but a state of perennial emulsion, of input-output and controlled chain reactions, such as exists in atomic simulation chambers. The "energy" of the mass must be liberated in order to be transformed into the "social".

It is a contradictory process, however, because information, in all its forms, instead of intensifying or even creating the "social relationship" is, on the contrary, an entropic process, a modality of the extinction of the social.

The idea is that the masses are structured, their captive social energy liberated, by the injection of information and messages (it is not so much the circumscribing action of the institutions, as the quantity of information received and the rate of exposure to the media which is the measure of socialization today). But the opposite is true. Instead of transforming the mass into energy, information simply produces more and more mass. Instead of informing, as it claims, that is to say, conferring form and structure, information increasingly neutralizes the "social field", creating ever larger inert masses, impervious to classic social institutions, to the very content of information itself. The nuclear explosion of symbolic structures by the social, and its rational violence, is succeeded today, by the fission of the social itself, by the irrational violence of the media and information — the final result being precisely an atomized, nuclearized, molecularized mass.

All that is left are fluid, mute masses, the variable equations of surveys, objects of perpetual tests in which, as in an acid solution, they are dissolved. Testing, probing, contacting, soliciting, informing — these are tactics of microbiological warfare undermining the social by infinitesimal dissuasion, so that there is no longer even time enough for crystallization to take place. Hitherto, violence used to crystallize the social, violently bringing forth an antagonistic social energy: a repressive demiurgy. Today it is a gentle *semi*urgy which controls us.

Thus information dissolves meaning and the social into a sort of nebula and is committed, not to increased innovation, but on the contrary, to total entropy.

I have so far dealt with information only in the social sphere of communication. It would be interesting, however, to carry the hypothesis into the domain of cybernetic information theory. There too, the obtaining thesis holds that cybernetic information is synonymous with negentropy, resistance to entropy, increased meaning, and improved organization. But it behoves us to pose the inverse hypothesis: Information = Entropy. For example: the information or knowledge which it is possible to have about a system or an event is already a form of neutralization and entropy of that system. This is true of the sciences in general and of the human and social sciences in particular. The information which reflects or diffuses an event is already a degraded form of that event. The role of the media in May, 1968 is a case in point. The coverage given to the student revolt gave rise to the general strike, but the latter turned out to be precisely a neutralizing black box, an antidote to the initial virulence of the movement. The publicity itself was a death trap. One must be wary of the attempt to universalize strategies through the dissemination of information, suspicious of all-out campaigns for a solidarity which is both electronic and fashionable. Every strategy which is geared to the universalization of differences is an entropic strategy of the system.

Mythinformation

LANGDON WINNER

Mythinformation (n.): The almost religious conviction that a widespread adoption of computers and communications systems, along with broad access to electronic information, will automatically produce a better world for humanity.

The specter of computer revolution is haunting modern society. Books, magazine articles, and news-media specials declare that this upheaval is underway, that nothing will escape unchanged. Like political revolutionists, advocates of computerization believe that a glorious transformation is sweeping the world and that they are its vanguard.

Of course, modern society has long since gotten used to "revolutions" in laundry detergents, underarm deodorants, floor waxes, and other consumer products. Exhausted in advertising slogans, the revolution image has lost much of its punch. Those who employ it to talk about computers and society, however, appear to make much more serious claims.

According to visionaries like Edward A. Feigenbaum and Pamela McCorduck (*The Fifth Generation*) or Murray Turoff and Starr Roxanne Hiltz (*The Network Nation*) industrial society, which depends on material production for its livelihood, is being supplanted by a society in which information services will enable people to satisfy their economic and social needs. As computation and communication technologies become less expensive and more convenient, all the people of the world, not just the wealthy, will use the wonderful services that information machines make available. Gradually, existing differences between rich and poor will evaporate.

Long lists of services are meant to suggest the coming utopia: interactive television, electronic funds transfer, computer-aided instruction, customized news service, electronic magazines, electronic mail, computer teleconferencing, on-line stock and weather reports, computerized yellow pages, shopping via home computer, and so forth. In the words of James Martin, writing in *Telematic Society*: "The electronic revolution will not do away with work, but it does hold out some promises: most boring jobs can be done by machines; lengthy commuting can be avoided; the opportunities for personal creativity will be unlimited."

In this interpretation, the prospects for participatory democracy have never been brighter, offering all the democratic benefits of the ancient Greek city-state, the Israeli kibbutz, and the New England town meeting. J. C. R. Licklider, a computer scientist at MIT, writes hopefully in a 1980 article called "Computers and Government": "The political process would essentially be a giant teleconference, and a campaign would be a months-long series of communications among candidates, propagandists, commentators, political action groups, and voters. The information revolution is bringing with it a key that may open the door to a new era of involvement and participation."

Mythinformation in the High-tech Era

Taken as a whole, beliefs like these make up what I call *mythinformation*: the almost religious conviction that a widespread adoption of computers and communications systems, along with broad access to electronic information, will automatically produce a better world for humanity.

It is common for the advent of a new technology to provide occasion for flights of utopian fancy. During the last two centuries the factory system, railroads, the telephone, electricity, automobiles, airplanes, radio, television, and nuclear power have all figured prominently in the belief that a new and glorious age was about to begin. But even within the great tradition of optimistic technophilia, current dreams of a "computer age" stand out as exaggerated and unrealistic. Because they have such broad appeal, and because they overshadow other ways of looking at the matter, these notions deserve closer inspection.

As is generally true of myths, the dreams contain elements of truth. What were once industrial societies are being transformed into service economies, a trend that emerges as a greater share of material production shifts to the developing countries, where labor costs are low and business tax breaks are lucrative. However, this shift does not mean that future employment possibilities will flow largely from the microelectronics and information-services industries, even though some service industries do depend on highly sophisticated computer and communications systems.

A number of studies, including those of the US Bureau of Labor Statistics, suggest that the vast majority of new jobs will be menial service positions paying relatively low wages. As robots and computer software absorb an increasing share of factory and office tasks, the "information society" will offer plenty of work for janitors, hospital orderlies, and fast-food helpers.

The computer savants correctly notice that computerization alters relationships of social power and control; however, the most obvious beneficiaries of this change are large transnational business corporations. While their "global reach" does not arise solely from the application of information technologies, such organizations are uniquely situated to exploit the new electronic possibilities for greater efficiency, productivity, command, and control. Other notable beneficiaries will be public bureaucracies, intelligence agencies, and ever-expanding military organizations.

Ordinary people are, of course, strongly affected by these organizations and by the rapid spread of new electronic systems in banking, insurance, taxation, work, home entertainment, and the like. They are counted on to be eventual eager buyers of hardware, software, and communications services.

But where is any motion toward increased democratization and social equality, or the dawn of a cultural renaissance? Current empirical studies of computers and social change — such as those described in *Computers and Politics* by James Danzig — suggest an increase in power by those who already have a great deal of power, an enhanced centralization of control by those already in control, and an augmentation of wealth by the already wealthy. If there is to be a computer revolution, it will most likely have a distinctly conservative character.

Granted, such prominent trends could be altered. A society strongly rooted in computer and telecommunications systems could incorporate participatory democracy, decentralized control, and social equality. However, such progress would involve concerted efforts to remove the many difficult obstacles blocking those ends, and the writings of computer enthusiasts seldom propose such deliberate action. Instead, they suggest that the good society will be a natural spin-off from the proliferation of computing devices. They evidently assume no need to place limits upon concentrations of power in the information age.

There is nothing new in this assumption. Computer romanticism strongly resembles a common nineteenth- and twentieth-century faith that expects to generate freedom, democracy, and justice through simple material abundance. From that point of view, there is no need for serious inquiry into the appropriate design of new institutions for the distribution of rewards and burdens. In previous versions of this conviction, the abundant (and therefore democratic) world would be found in a limitless supply of houses and consumer goods. Now "access to information" has moved to the top of the list.

Probing the Key Assumptions

The political arguments of computer romantics draw upon four key assumptions: 1) people are bereft of information; 2) information is knowledge; 3) knowledge is power; and 4) increased access to information enhances democracy and equalizes social power.

1. Is it true that people face serious shortages of information? To read the literature on the computer revolution, one would suppose this to be a problem on a par with the energy crisis of the 1970s. The persuasiveness of this notion borrows from our sense that literacy, education, knowledge, well-informed minds, and the widespread availability of tools of inquiry are of unquestionable social value.

Alas, the idea is entirely faulty. It mistakes sheer supply of information for an educated ability to gain knowledge and act effectively. Even highly developed societies contain chronic inequalities in the distribution of education and intellectual skills. The US Army must reject many of the young men and women it recruits because they cannot read military manuals.

If the solution to problems of illiteracy and poor education were a question of information supply alone, then the best policy might be to increase the number of well-stocked libraries, especially in places where libraries do not presently exist. Of course, that would do little good unless people were sufficiently well educated to use those libraries. Computer enthusiasts, however, are not known for their support of public libraries and schools; they call for *electronic information* carried by *networks*. To look to those instruments first while ignoring everything history has taught us about how to educate and stimulate a human mind is grave foolishness.

2. What is the "information" so cherished as knowledge? It is not understanding, enlightenment, critical thought, timeless wisdom, or the content of a well-educated mind. Looking closely at the writings of computer enthusiasts, "information" means enormous quantities

of data manipulated by various kinds of electronic media, used to facilitate the transactions of large, complex organizations. In this context, the sheer quantity of information presents a formidable challenge. Modern organizations continually face "overload", a flood of data that threatens to become unintelligible. Computers provide one way to confront that problem; speed conquers quantity.

The information most crucial to modern organizations is highly time-specific. Data on stock market prices, airline traffic, weather conditions, international economic indicators, military intelligence, and public opinion polls are useful for very short periods of time.

Systems that gather, organize, analyze, and use electronic data must be closely tuned to the latest developments. Information is a perishable commodity.

But is it sensible to transfer this ideology, as many evidently wish, to all parts of human life? A recent *Business Week* article on home computers concluded: "Running a household is actually like running a small business. You have to worry about inventory control — of household supplies — and budgeting for school tuition, housekeepers' salaries, and all the rest." One begins to wonder how running a home was possible before microelectronics.

3. "As everybody knows, knowledge is power," wrote Dr. Feigenbaum. This attractive idea is highly misleading. Knowledge employed in particular circumstances may well help one act effectively — a citrus farmer's knowledge of frost conditions enables him* to fight the harmful effects of cold snaps. But there is no automatic, positive link between knowledge and power, especially power in a social or political sense. At times, knowledge brings merely an enlightened impotence or paralysis. What conditions might enable ordinary folks to translate their knowledge into renewed power? It is a question computer enthusiasts ought to explore.

4. An equally serious misconception among computer enthusiasts is the belief that democracy is largely a matter of distributing information. This assertion plays on the valid beliefs that a democratic public should be open-minded and well-informed, and that totalitarian societies are evil because they dictate what people can know and impose secrecy to restrict freedom. But democracy is not founded primarily upon the availability of information. It is distinguished from other political forms by the recognition that the people as a whole are capable of, and have the right to, self-government.

There are many reasons why relatively low levels of citizen

*See "Questioning Technology, Questioning Patriarchy," p. 1.

participation prevail in some modern democracies, including the United States. Perhaps opportunities to serve in a public office or influence policy are too limited; in that case, broaden the opportunities. Or perhaps choices placed before citizens are so pallid that boredom is a valid response; then improve the quality of those choices. But it is not reasonable to assume that a universal grid of sophisticated information machines, in itself, would stimulate a renewed sense of political involvement and participation.

The role of television in modern politics suggests why this is so. Public participation in voting has steadily declined as television replaces the face-to-face politics of precincts and neighborhoods. The passive monitoring of electronic news makes citizens feel involved while releasing them from the desire to take an active part, and from any genuine political knowledge based on first-hand experience. The vitality of democratic politics depends on people's willingness to act together — to appear before each other in person, speak their minds, deliberate, and decide what they will do. This is considerably different from the model upheld as a breakthrough for democracy: logging onto one's computer, receiving the latest information, and sending back a digitized response. No computer enthusiasm is more poignant than the faith that the personal computer, as it becomes more sophisticated, cheaper, and more simple to use, will become a potent equalizer in society. Presumably, ordinary citizens equipped with microcomputers will counter the influence of large, computer-based organizations. This notion echoes the eighteenth- and nineteenth-century revolutionary belief that placing firearms in the hands of the people would overthrow entrenched authority. But the military defeat of the Paris Commune in 1871 made clear that arming the people may not be enough. Using a personal computer makes one no more powerful vis-a-vis, say, the US National Security Agency than flying a hang glider establishes a person as a match for the US Air Force.

The Long-term Consequences

If the long-term consequences of computerization are anything like the ones commonly predicted, they will require rethinking of many fundamental conditions and institutions in social and political life. Three areas of concern seem paramount. First, as people handle more of their daily activites electronically — mail, banking, shopping, entertainment, travel plans, and so on — it becomes technically feasible to monitor these activities with unprecedented ease. An age rich in electronic information may achieve wonderful social conveniences at the cost of placing freedom — and the feeling of freedom — in a deep chill.

Second, a computerized world will renovate conditions of human sociability. Indeed, the point of many applications of microelectronics is to eliminate social layers that were previously needed. Computerized bank tellers have largely done away with local branch banks, which were places where people met and socialized. The so-called electronic cottage would operate well without the human interaction that characterizes office work.

These developments pare away the face-to-face contact that once provided buffers between individuals and organized power. Workers who might previously have recognized a common grievance and acted together to remedy it are now deprived of such contact, and thus increasingly influenced by employers, news media, advertisers, and national political leaders. Where will we find new institutions to balance and mediate such power?

Third, computers, satellites, and telecommunications may recast the basic structure of political order, as they fulfill the modern dream of conquering space and time. These systems make possible instantaneous action anywhere on the globe without limits imposed by the location of the initiator. But humans and their societies have traditionally lived, acted, and found meaning within spatial and temporal limits. Microelectronics tends to dissolve these limits, thereby threatening the integrity of social and political forms that depend on them.

Transnational corporations of enormous size can now manage their activities efficiently across the whole surface of the planet. If it seems convenient, operations can be shifted from Sunnyvale to Singapore at the flick of a switch. In recent past, corporations have had to demonstrate at least some semblance of commitment to their geographical base; their public relations often stressed the fact that they were "good neighbors". But when organizations are located everywhere and nowhere this commitment easily evaporates. Towns, cities, regions, and whole nations must swallow their pride and negotiate for favors. Political authority is gradually redefined.

By calling the changes of computerization "revolutionary", people tacitly acknowledge that these changes require reflection; they may even require strong public action to ensure desirable outcome. Yet the occasions in our society for reflection, debate and public check are rare indeed. The important decisions are left in private hands inspired by narrowly focused economic motives. While it is widely recognized that these decisions have profound cumulative consequence for our common lives few seem prepared to own up to that fact. Some observers forecast that the computer revolution will be guided by new wonders in artificial intelligence. Its present course is influenced by something much more familiar; the absent mind.

Who Knows: Information in the Age of the Fortune 500

HERBERT I. SCHILLER

Technology plays a vital role in the emerging new scheme of things. It serves dually: first, to integrate the transnational corporate system, and second, to deepen the dependence of the peripheral world on hardware, software, training, and administration supplied by that system.

The less developed nations are not to be denied the new technology. On the contrary, technology is being pressed on the poorer countries in an atmosphere of urgency. "We must offer to expand communication systems abroad," urges one promoter of US information policy. "Imaginative use of our satellites and earth stations, shared time on our broadcasting channels, crash projects to produce cheap newsprint — all and more are readily possible."[1]

Technology's role in the less-developed economies will be extended, but under the auspices of the transnational corporate system. This, it is reasonable to believe, is intended to assure the implantation of Western developmental models — of production, administration, consumption, and education. Though it is not likely that the *most* sophisticated instrumentation and processes of advanced capitalism will be made available to the peripheral world, even in those cases where it is offered the effects will be the same — dependency, and development patterned on the market model. Writing about the possible transfer of information technology, one observer notes:

Even if the US Government did subsidize access to US data banks and "information resources" [the US proposal at the UNESCO 1978 Paris meeeting and elsewhere], it is questionable how useful this information would be .... In order for information to have real utility it has to be tailored to the needs and circumstances of the user, and this simply is not achieved through the installation of an international data network. However, once the technical, financial, and management skills and infrastructures have been developed, as in the case of Taiwan, Hong Kong, Malaysia and others, then the export of information is related to the export of capability and "comparative advantage"[2]

The important consideration at this time, from the perspective of the transnational corporate policymakers, is to get advanced communications technology installed in as many places as quickly as possible. The effort to put the technical infrastructure in place — termed "operationalism"[3] — is as agreeable to the American information industry complex as it is to the transnational system overall.

Suggestive of the efforts undertaken to create an international atmosphere of encouragement, if not urgency, for the rapid adoption of new communications technology, the International Telecommunications Union (ITU) organized a forum in Geneva in September 1979, the introductory section of which was called Telecommunications Perspectives and Economic Implications. The subjects under discussion were: "strategies for dealing with evolving international telecommunications; industrial products and transfer of technology for effective operation; telecommunication services and networks; [and] financing of telecommunications".

Offering views on these important questions was a panel of speakers recruited almost exclusively from the most powerful companies producing equipment in the transnational corporate system. Among them were: the president of RCA, the Vice-President and Chief Scientist of IBM, the President of Siemens AG, the Vice-President of the Executive Board at Philips; the Executive Vice-President of A.T.&T.; the Vice-President and Group Executive of Hughes Aircraft, and similar ranking officers from Thomson-CSF, NASA, Comsat, and Ericsson (Sweden). Supplying the new instrumentation and processes means consolidating Western long term control in international markets over equipment, replacement parts, servicing, and finance.

"Operationalism", consequently, makes little attempt to ascertain the appropriateness of the items sold or to develop norms for such an evaluation. Useful or not, needed or not, Western suppliers push their wares come what may. Mild attempts to establish international specifications and protective standards are rebuffed and labelled "premature".[4]

Information Inequality in the Center of the System

Though this is small consolation to the rest of the world, in the global shift of economic and informational activity now proceeding *the center of the system, no less than the periphery, experiences deepening inequalities.*

To the developing nations, the new communications technology is promoted as a means of lessening social gaps in education and literacy and as a means of leapfrogging into the modern age, with classrooms and businesses informed from satellite broadcasts. For the already industrialized countries, the promise is of electronically induced democracy, plebiscites, and polls, carried out at home by the touch of a button on the living room TV console.

These are claims the transnational corporate system circulates through its transnational media circuits. Actual developments, in the United States and elsewhere, present another reality. Consistent with market practice, the advantages and capability of using the new instrumentation vary directly with technical and financial ability (to pay). It requires little imagination to guess who benefits from the new information technology in corporation-dominated America, where a few hundred businesses control more than three-fifths of the national economy.

In the very heart of the most advanced "information society" it is predicted that, contrary to widely publicized claims, the introduction of electronic information systems will deepen information inequality in the social order.[5] And if this is the prospect for a nation with an abundance of informational circuits, what may be expected to occur in those numerous countries where scarcity, weakness, and dependency are still the prevailing characteristics?

The Free Flow of Information

In the period from 1940 to 1970, "the free flow of information" was a major element in United States information and foreign policy.[6] In this earlier time, the objectives generally were limited to the market needs of the American media industries — the news agencies, film and TV-program export sectors, advertising, publishing, and record industries. The doctrine also was directed against the socialist nations, intended to put pressure on them to open up and be susceptible to consumerist ideology.

Though these aims remain important objectives, the information issue, and the free flow of information doctrine in particular, now transcend the somewhat parochial question of expanded markets for American media interests. Information gathering, processing and transmission, have become essential and determining elements in affecting corporate America's position in a new international economic order. William Colby, former director of the Central Intelligence Agency, puts it this way:

Today the world is facing a choice between free trade and protectionism in international information exchange. As we learned in the field of commodities, it will be important to choose the path toward freedom rather than protectionism . . . . In the information industry, whether hardware, software, or the rapidly growing field of substantive analysis, a similar strategy for free international exchange must be developed. The benefits of free exchange and the cost of attempts to obstruct it must be articulated.[7]

Just how important the free flow of information doctrine has become to the transnational corporate system's maintenance and survival is noted by another communications analyst. It is, Clippenger writes, "the pillar not only of US civil liberties and individual freedom, but the market economy as well . . . . And many US commentators and officials have avowed that the concept of free flow is a wholly non-negotiable item.[8] If supremacy in the information sector cannot be maintained, American power could shrink to continental confines. The capability to monitor and police the world, for the protection of the corporate system, is also at stake. Former President Carter alluded to this capacity in discussing the SALT II Treaty with Congress:

As I have said many times, SALT II is not based on trust. Compliance will be assured by our own nation's means of verification, including extremely sophisticated satellites, powerful electronic systems, and a vast intelligence network.[9]

Similarly, the Director of Security Programs for IBM — the corporate empire of computing capability — mentions circumspectly the communication functions vital to the survival of the worldwide American business system:

I believe the paramount objective should be the preservation of the free flow of information across and within national borders balanced by considerations for privacy and national security. This must be the single most important objective simply because of the immense economic and political impact of free information flow in today's society. So much of this information flow takes place beneath the surface of our conscious activities that we literally take it for granted. Perhaps we would only realize its true value and impact if it were restricted.[10]

But historically and currently, the free flow of information is a myth. Selectors and controllers continue, as they always have, to sift and shape the messages that circulate in society. It is always a matter of who the selectors are and whom they represent. And this is an area of which social class is in control.

The free flow of information, as a phrase, does describe well, however, what now occurs in the global information infrastructures that link the transnational business system. In these *privately* organized circuits, the flow *does* move freely between corporate affiliates in the international sphere. Public scrutiny is avoided and strong efforts are made to keep matters as they are.*

But the *Fortune 500* companies are not necessarily limited to transmitting data within their corporate structures. As new technology makes all communications processes increasingly interchangeable — i.e., messages, whatever the form, be it record, voice or visual — are reducible to electric impulses. Transnational businesses have the opportunity to reach large audiences and publics on their own terms, possibly through their own informational circuits.

A recent study notes: "The ability to communicate with masses of

---

* Some indication of the intra-transnational corporate information flow internationally is provided by Hewlett-Packard, a major US manufacturing company: "Hewlett-Packard manufactures more than 4000 products for wide-ranging markets which are primarily in manufacturing-related industries. We have 38 manufacturing facilities and 172 sales and service offices around the world, and together these employ about 45,000 people. We have experienced a very rapid growth of about 20% per year, culminating in sales of $1.7billion in 1978 . . . . To support this business, we currently have some 1,400 computers (not including desktop units or handheld calculators). Of these, 85% are used to support engineering and production applications, are usually dedicated to specific tasks, and often are arranged in networks. A number of them are also used in computer-aided design applications as front-end processors for large mainframes. The remaining 200 computers are used to support business applications . . . . The network tying all this together consists of 110 data communications facilities located at sales and service offices, at manufacturing plants, and at corporate offices in Northern California and Switzerland." (Cort Van Rensselaer, "Centralize? Decentralize? Distribute?", *Datamation*, April 1979, 25 (4), p.90).

people is spreading beyond the 'institutional media'." And a United States Supreme Court decision in 1978 approved the principle that "a telephone company, or any other corporation, has First Amendment rights".[11]

In recent years, United States policy makers defended American broadcasters' rights to use direct satellite broadcasting — when it became available — without complying with national oversight from any country. The argument advanced claimed that the world itself was covered by the US Constitution. Now, apparently expanding on that modest interpretation, the international community may be informed that not only American media but *all* US transnational corporations have unlimited global communication rights because they are shielded by the US Bill of Rights.[12]

1 Leonard R. Sussman, "A new world information order?" in *Freedom at Issue*, November/December 1978 (48), p. 9.

2 John H. Clippenger, "The hidden agenda" in *Journal of Communication*, Winter 1979, 29 (1), pp. 189-190.

3 Benno Signitzer, *Regulation of Direct Broadcasting from Satellites: The UN Involvement* (New York: Praeger, 1976).

4 The US delegate to the Strategies and Policies for Informatics (SPIN) meeting in Torremolinos, Spain, August 1978, stated this explicitly.

5 Herbert S. Dordick, Helen G. Bradley, Burt Nanus and Thomas H. Martin, "The Emerging Network Marketplace" F35, December 1978. Center for Futures Research, Graduate School of Business Administration, University of Southern California, Los Angeles, CA 90007.

6 Herbert I. Schiller, *Communication and Cultural Domination*, M. E. Sharpe, White Plains, New York, 1976, Chapter 2.

7 William E. Colby, "International Information — Free Trade or Protectionism", International Conference on Transnational Data Flows, Washington DC, 3 December 1979.

8 John H. Clippenger, *op. cit.*, p. 199.

9 *The New York Times*, 19 June 1979, p. A-13.

10 Harry B. DeMaio, "Transnational Information Flow: A Perspective", *Data Regulation: European and Third World Realities* (Uxbridge, England: Online, 1978), p. 170.

11 William H. Read, "The First Amendment Meets the Second Revolution", Working Papers, W-79-3. Harvard University Program on Information Resources Policy, Cambridge, Massachusetts, March 1979, pp. 25-26.

12 A Canadian Government study sees this as an imminent reality. "Few people in Canada are aware of the implications of what is happening. These are some of the dangers foreseen if protective measures are not urgently devised and implemented. Greater use of foreign, mainly US, computing services and growing dependence on them will . . . facilitate the attempts of the government of the United States to make laws applicable outside US territory." *Telecommunications and Canada* (The Clyne Report), p. 64.

*9. How — and how effectively — is the technological
outcome of science regulated? Is there some research and
development that should be off-limits?*

In a recent series of radio interviews of nationally recognized
research scientists, this question was asked of each person:
"Who is responsible for the technological consequences of
scientific research?" "The engineers," said one scientist. "The
public", said another. "Not the scientists", stated a third
savant. The others said they didn't know. Not one of them
ventured to suggest that the *scientist* might have responsibility.

The technician's answer to the culpability question could be
Eichmann's: "I was only following orders." As for the public:
"What do we know? It's probably for the best. It's inevitable
— you can't stop progress." (That's what people said about
slavery, among other scourges.)

Eugene Schwartz takes the measure of the juggernaut of
technology, and concludes that the prevailing antidote to high
tech problems — more technology — is not the answer. Greg
Davis shows the weaknesses in sanguine assumptions about
regulation of technology, and discusses some grass roots
combat strategies. Leonard Cole is cautiously optimistic as to
prospects for successful control of science and technology,
whereas T. Fulano attacks faith in science as the triumph of
the inhuman and the death of the wonder of nature.

Overskill: The Decline of Technology in Modern Civilization

EUGENE S. SCHWARTZ

A wall of water gushes from a fountain on the campus of the Illinois Institute of Technology in Chicago in a courtyard bounded by the austere, rigidly functional buildings of Mies van der Rohe. The water is forced four feet into the air in the shape of an inverted cone to form a chrysanthemum pattern before falling back into a pool below. Pressures, temperatures, and flow rates have been calculated to spew water from a specially designed circular spout that circulates it throughout the year. A passerby in subzero winter can witness the strange sight of water spouting and falling back on ice and snow encrusting the fountain.

No one lingers about the fountain in the winter, nor does it attract watchers in the warmer seasons. Students do not tarry to gaze at the fountain because its solid wall of rising water connotes power, not contemplation. No gentle spray rises from the spout to catch sunbeams and convert them into rainbows. There is no trickling or dripping or gurgling, only a ponderous rising and falling back. The fountain is too small to imitate the beauty of Chicago's Buckingham Fountain, several miles distant on the lakefront; it is too gross to capture the mood of the nearby Dove Girl and Turtle Boy fountains.

The fountain spouts forth in summer heat and winter cold, unattended, unwatched — a symbol of a technology unrelated to human purpose and human aspirations.

Such mindless technology is inundating the world, sweeping all before it. Although the graphs of "progress" continue to rise, man is only now beginning to perceive the price that is exacted to enable

* See "Questioning Technology, Questioning Patriarchy," p.1.

the fountains of technology to continue to spew forth their promise. Mingled with the "Eureka" cries that accompany the scaling of new heights of scientific achievement are the rising wails of mankind who suffer from these same achievements. An adventurous and truth-seeking science is becoming an appendage of the technology it spawned; truth is becoming that which politics and economics in their parochial national manifestations demand; freedom of inquiry and revolt against authority are being transformed into a new orthodoxy administered and controlled by a new elite. Science and technology have become established myths of the descendants of those forebears who sought to abolish myths in the name of reason. Science has become a secular religion; technology is its temple, efficiency its dogma.

What if the religion of technology is not a truth but a falsification of man and nature and society? What if this religion will bring about a homogenization of all cultures so as to preclude multiple responses to the dangers of the present and the challenges of the future? What if the elaborate mythology erected over the past three hundred years and reaching maturity in our era is found to be spun from the same figments of self-deception, unrequited wishes, and dreams that have characterized men's myths from the dawn of human life on earth?

It does not matter that the new mythology is shrouded in the cloak of science with its purity and vision attended by an esoteric priesthood of savants and corporate technicians. Their Pythagorean mystique is enshrined in huge temples built to worship the atom and the machine. They are the guardians of the cabbala that guides space ships, plots the trajectories of hydrogen warheads, and directs the "efficient" operation of the industrial economy. They communicate with the gods of automation who dwell in the wired intricacies of computers. They count and recite their symbolic litanies inherited from the notched sticks and abacus of the Orient, Napier's bones, and the calculators of Pascal, Leibniz, and Babbage.

The philosophy of "progress", although ostensibly based on the "reason" of science, was an act of faith, a belief in a secular religion. It supplanted earlier mythologies embedded in magic from which science sprang. Science triumphed over a less reasonable magic, because, pragmatically, science proved to be more efficient than magic. After all, were not the Whitsuntide festivals and sexual couplings in the fields of Peru and Java and West Africa to promote fertility of the soil and bountiful crops dispensed with because the application of trisodium phosphate to the soil resulted in increased yields of crops?

Technology became the servant of science in translating scientific "truths" into better life for the peoples of the earth. The secular religion was hailed as the new Messiah that would bring {man} surcease from toil and freedom from want. It would liberate the enslaved peoples of the earth; it would feed the hungry and cure the sick. It would carry{man}forward to progress, satiety, power.

The steel mill, the atomic reactor, one's country's name emblazoned on an orbiting satellite — these were the symbols in the dreams of leaders of Third World nations.* An infrastructure of roads, communications, and factories — this was the promise of the "haves" to the "have-nots". Scientists and technicians invaded the developing nations, proselytizing for the new religion.

The religion of progress through science and technology combined elements of exaltation and grandeur with the arrogance of established religions. Mortal{man} has escaped earth's gravitation and trod upon the moon. Through biological alchemy, the promise of immortality was held to be within reach. Scientific{man}has no need for that other God when the universe is opened by rockets and the promise of immortality is made by genetics and molecular biology. The fire and brimstone of hell can no longer terrorize{man} who has experienced Dachau and Hiroshima. God, as a cultural anachronism, appears quaintly in an astronaut's prayers and is invoked to bless bombs and soldiers, but science and technology direct and power{man's}restless spirit.

In the historically short time span of three hundred years, science and technology have transformed the earth and its inhabitants. The earth pulses and labors. The electron and the atom have become genii to reshape the earth. Time is condensed, space telescoped. The word approaches a technological unity. The Messiah is nigh.

Alas! Even as the vision becomes intensified, the reality recedes. "The Road we have been travelling is deceptively easy, a smooth superhighway on which we progress with great speed, but at its end lies disaster," Rachel Carson warned in *Silent Spring*.[1]{Mankind,}

---

* The "Third World" refers, generally, to the pre-technological societies located in Central and South America and in Africa and Asia. The "First" and "Second" worlds (we will leave it to future historians to ascertain the order) refer, generally, to the technological societies of the capitalist and communist countries. "Third World" is used instead of "underdeveloped" or "developing nations" because the latter terms are normative in the sense that the technologically developed nations are considered to be the desired norm in which all nations must be compared. This dichtomization is part of the arrogance of the technological societies. In a humanistic sense, the developed nations are becoming increasingly underdeveloped.

meanwhile, ponders: We have sowed our fields with trisodium phosphate and there are more hungry people in the world than ever before. We have built hospitals and clinics and there are more sick people than ever before. We have built schools and illiteracy flourishes. We have built factories and filled them with machines and find that we are slaves to the machines. We have cut down our forests, depleted our natural resources, and overmined the earth. We have poisoned our lakes and streams, polluted the air we breathe, and transformed the face of the earth into a labyrinth of concrete ribbons, scarred hills, and monstrous slag heaps.

Living space is compressed on a crowded planet. We have cleared the jungles to build cities and have transformed our cities into jungles. We have erected skyscrapers to house millions, and we pass each other in the corridors as strangers. We have spewed out automobiles by the millions, and they crawl on the congested highways at a pace slower than a horse-drawn carriage. Speedy jet planes carry {man} back and forth across the earth as {he} seeks to escape problems at home and finds that they have taken root in foreign lands. The rich grow richer, the poor poorer. The satiated grow sick from overconsumption; the hungry sicken from starvation. The technological machine continues to disgorge products {man} neither needs nor wants.

The despoliation of the earth and the poisoning of {man's} habitat go hand in hand with the increasing anarchy between nations, the increasing breakdown of social institutions, and the growing alienation of individual men. Civilization is increasingly becoming the neurosis about which Freud wrote with great pessimism.

The issue is not one of technological optimism or technological pessimism. Posing the problem in this form begs the question. The problem is one of human survival. The role technology can play in the struggle to retain human life on the planet is one of the problems {mankind} now faces. "It seems possible that the new amount of technological power let loose in an overcrowded world may overload any system we might devise for its control; the possibility of a complete and apocalyptic end of civilization cannot be dismissed as a morbid fantasy."[2] This statement by Don K. Price, former president of the American Association for the Advancement of Science, calls for a new technological realism. The drama now being played out on planet earth does not brook any facile simplifications of a Janus-faced technology that has the potential for both good and evil. Wise {men} will, it is implied, accentuate the good; venal or

·ignorant⸨men⸩will choose the evil. Technology as a human-devised means to solve problems is itself in question.

Belatedly, Western ⸨man⸩ is beginning to realize what ⸨he⸩ has wrought on his limited habitat. The earth is a finite territory with finite resources. In wasting and despoiling this bounty, ⸨man⸩ is faced with limits — limits in space, in food, in raw materials, in pure water, in fresh air. The ravaged earth does not fight back but presents its stricken face to a civilization that only now is dimly becoming aware of the limits on the possible and can acknowledge its hubris in nothing more than contrite despair and fear.

The violence against the earth and ⸨man⸩ that has been the mainspring of the Industrial Revolution has run amok as the possibilities for future civilizations begin to be foreclosed. Technology has hastened the process of foreclosure by homogenizing the world and further reducing alternative futures before the peoples of the earth. With its shibboleth of efficiency, "progress" has jeopardized human survival, for nature through countless millennia of evolution has never been efficient. Survival has been achieved through safety, through maintenance of reserves, through following myriads of paths, through exploration of many potentials, through proliferation of species, through unpredictable divergences and mutations, both biological and social.

What has gone wrong with the "progress" that has led ⸨man⸩ in but a short three hundred years to the edge of disaster? Has frail ⸨man⸩ with a propensity for both good and evil misused "reason"? Have the "relations of production" in private hands subverted the utopia that might have been? Has science for too long been influenced by a reductionist technique whereby it is assumed that breaking everything down into its smallest parts will reveal the whole, whereas a holistic science of interdisciplinary endeavors will make the necessary adjustments? Has piecemeal "progress" been· too fragmented, whereas a cybernetic model of an integrated "system" will lead to more efficient planning and control? If ⸨man⸩ has unlocked the secret of the atom and trod upon the moon, cannot similar crash programs, through the mobilization of money and scientist-technicians, solve the problems that endanger ⸨man⸩?

It is the thesis of this book that the answer to the above questions is uniformly "no". ⸨Man⸩ has not misused the "reason" of science and technology to bring us to our present state. Instead, the tragedy of the present was inherent in the basic premises of science from its early formulations of the modern age, beginning with Galileo Galilei, Francis Bacon, and René Descartes, and no reformulation of the questions with which science deals will alter these philosophical defects.

What then of technology? There is a school of thought that says all technology is bad. It is a destroyer of all human values; it is autonomous and has become an end in itself. Another school, the predominant one, states that technology is a great blessing, powering "progress" and advancing humanity while improving man's condition as predicted by the Enlightenment prophets. A third school maintains that technology is but the continuation of the advances that {mankind} has registered throughout the ages. It is nothing new, marks no revolutionary break with former practices, and is well recognized as a factor in social change.[3]

An increasing number of technicians, administrators, and scholars, however, confronted with the crises to which technology has contributed, are beginning to ask questions about the role of technology in society. Out of this questioning has come a range of suggested policies. There are those who speak of controlling science through law. The United States Constitution, for example, would add the right to a wholesome environment to the Declaration's rights "to life, liberty, and the pursuit of happiness". The futurists contend that by predicting possible futures on the basis of present and foreseen developments, the future can be controlled and policies, both private and governmental, can be undertaken to realize the most desirable futures. Technology assessment, its proponents claim, can analyze in advance the benefits and risks of exploiting new technological capabilities. Decisions would then be made by political and economic institutions to select those technologies that promise positive results, while those with deleterious results would be withheld. This latter policy is consistent with the views of those who urge technological renunciation or technological disarmament. {Man's} finger, in this view, is on an environmental trigger that can destroy {man} as surely as can nuclear weapons. The way to control is through abjuring those technological developments that may have short-range benefits but long-range defects.

The most vocal and widespread policy advocated to meet the crisis engendered by the revolutionary transformation of the earth and society by technology is — more technology. Writing in *Science*, the mouthpiece of the American scientific establishment, Professor Harvey Brooks of Harvard states the case for more technology to cure the ills technology has brought in its wake:

With respect to the great modern problems — what I call the four P's of population, pollution, peace, and poverty — it may be that articulating these is the most important part of the problem — that once these needs are formulated in the right way, the technological solutions will become obvious, or will fall into place.[4]

Can the reorientation of technology solve the problems technology has created? Can an extension of the scientific genius undo what the genius has done? Can more, larger, and more efficient techniques retrieve {man} from the catastrophes that threaten {him} and that arose from fewer, smaller, less efficient techniques?

Again the response is negative. New and more powerful techniques cannot solve the problems that technology has engendered because technology is a dialectical process arising from the relationships of {man's} interaction with nature. Technology is subject to change and conflict. It has limitations and constraints. It can also be self-destructive; it also undermines the science which supports it. The scientific enterprise and an expanded technology are likely to negate themselves and to extirpate {mankind} in the process.

Ours is an exhaustible world. We have lost much of our freedom to experiment and to choose because of the pressing urgency of the converging crises brought on, in the main, through the agency of the scientific and technological revolutions. We have lost space by shrinking the earth, and we have compressed time to the extent that a child grows up in a world that changes faster than {he} can adapt to it. If we rush pell-mell into a troubled and dangerous future with the same slogans and practices that have brought us to this situation, can we expect another disaster?

{Man's} massive and total assault against nature has been characterized by abysmal ignorance and monumental stupidity. The assault has been led by a political economy that has elevated greed, selfishness, and acquisitiveness to holy virtues and has been abetted by a science and technology that excluded all morality and ethics from its practices. No individual or group bore responsibility for the uses to which knowledge and praxis were put. It is as if the "invisible hand" of political economy were to be matched with an "invisible hand" that would mesh the disparate and diverging sciences and technologies into a human enterprise that would promote human welfare.

The "invisible hands" have brought unimagined wealth and comfort to a small fraction of the earth's population in the short space of three hundred years. To achieve this ephemeral end, the earth has been stripped of its resources and the human habitat has been made nearly uninhabitable. Faced with converging crises and the failure of the "invisible hands" on the economic and the ecological fronts, we are now told that the answer to our problems is "more of the same". More science, more technology, more research. The hubris that has brought man to the brink of catastrophe propels man on to accelerate that catastrophe.

Man, at the peak of a nebulous "progress", is threatened on one hand by a suffocating death and on the other hand by annihilating disintegration. We must now re-examine the tenets that brought us to this state. The reason that is a faith must be re-evaluated. The science that reached for the moon as mankind began to lose the earth must be questioned. The tools that harnessed nature but destroyed her in the process must be recast.

1. Rachel Carson, *Silent Spring* (Boston: Houghton Mifflin, 1962), p.277.
2. Don K. Price, "Purists and Politicians", *Science*, CLXIII (January 3, 1969), 25-31.
3. Emmanuel G. Mesthene, *Technological Change: Its Impact on Man and Society* (New York: New American Library, 1970), pp.15-19.
4. Harvey Brooks, "Applied Science and Technological Progress", *Science*, CLVI (June 30, 1967), 1712.

Technology — Humanism or Nihilism

GREGORY H. DAVIS

Faced with the possibility of controls, scientists will argue that they are the only ones who possess the necessary expertise to decide when controls should be established. Their specialization and expertise, however, are precisely why they must not be allowed to determine these matters alone. Scientists are trained to be objective — that is, to leave personal feelings and moral considerations out of their determinations. Besides, scientists' decisions and findings *qua* scientists are based primarily on quantitative factors. Why, then, are they the best qualified to make judgmental, qualitative decisions? As Barry Commoner suggests, the scientist's most appropriate role is to provide accurate and necessary information to those who are charged with the responsibility of making judgements about technological control.[1]

A panel or board of scientists should not be given the power to make decisions about technological controls on their own, although scientists could — and should — be included in a board with a broader composition. How should these boards be constituted? Certainly, those who have been educated to make qualitative judgments — philosophers, theologians, and humanists, for example — should be members of such regulatory agencies and panels, provided that they are made independent of external economic pressures. Responsible lay persons should also be assigned to such panels, as should independent scientists, including those, like ecologists, who do not have a reductionist perspective. In some cases, it might be desirable to include doctors and lawyers who have demonstrated enlightened concern for the public interest. The bill Senator Edward Kennedy drafted in 1977 (and eventually opposed, due to intensive lobbying by scientific researchers!)[2] to establish federal regulation of recombinant DNA (i.e. gene-grafting) research provided for the creation of a National Commission with eleven presidentially-appointed members, a majority of whom would not be scientists. A bill introduced into the California legislature the same year proposed inauguration of an eleven-person DNA Safety Commission whose members, appointed by the governor and legislative leaders, would include five lay persons, two scientists knowledgeable in the recombinant DNA research field, and four experts drawn from areas like law, medicine, social science, ethics, and human rights.[3]

What these examples suggest is that the problems of who will do the appointing, how members of various boards of technological control will be mixed, and what the tenure of personnel will be cannot be satisfactorily resolved by establishing one general formula. The answers depend rather on the specific technology in question. Appointees are more likely to be suitable, of course, if the person or persons doing the appointing are free from control by special interests and have a commitment to the values of Technological Humanism.

What about more democratic means of technological control? One possibility would be for citizens to elect the members of agencies and boards charged with the responsibility of technological control rather than having them appointed by government officials or other responsible citizens. This would be a step toward democracy in the formal sense, but the problem would then be the same as that which concerns the American electoral process itself. Unless adequate qualifications of education and experience were established, unless candidates were not economically connected with special interests, unless all candidates were given equal media time for live appearances, and unless the voters were informed about the issues, the results might be less satisfactory than selecting the panel members by appointment.

Another democratic possibility is favored by Barry Commoner. Let the people and their elected representatives themselves decide, says Commoner, after open, public debate about technological issues where scientists have met their responsibility for providing reliable information about risks and benefits.[4] It would be preferable, of course, to let the people themselves decide, since this eliminates the problem of elected representatives who disregard the public interest because they have been influenced by lobbyists or campaign contributions from special interests.

Unfortunately, there are difficulties with the popular decision approach, too. The problem is not, as many technocrat managers and scientific "experts" maintain, that the people won't understand the costs and benefits properly. They will understand, if scientists do their job of explaining responsibly and conscientiously. A difficulty will arise, however, if the economy is organized so that many people are dependent on the technology in question for their livelihood or think it may create future job opportunities for large numbers of individuals. Then short-term economic considerations may prevail over long-term survival and humanistic ones, no matter how compelling. Obviously, popular decisions on technological control will be more responsible in a society where the kind of self-sufficiency and decentralization promoted by the use of soft and appropriate technologies already exists.

Another difficulty will arise if voters are not adequately informed about the real issues and are deliberately kept in ignorance by the experts and are manipulated by media propaganda. Then they may make a wrong decision, although it will, indeed, be a democratic one. This is precisely what happened in California in 1976, when citizens voted on Proposition 15, a ballot measure to establish stringent safety requirements for nuclear power plants. The nuclear industry, which includes reactor manufacturers, construction and operator unions, energy companies who own uranium mines and milling facilities, and public utilities, spent a whopping $5·3 million for media propaganda[5] designed to convince California's passage of the measure would cause severe economic dislocations. Supporters of the proposition, mainly average middle-class citizens who donated $485,000 for their own media campaign[6] were outspent by a factor of eleven to one! The electorate, basically uninformed about the dangers of nuclear power because of years of government and industry dissimulation and misinformation, predictably succumbed to the nuclear industry's media blitz. The measure lost by a two-to-one margin, as the people were manipulated to vote against themselves.

At present, the indications are that the US Government, with its policy of "reindustrialization" developed under the Carter Administration and even more vigorously embraced by the Reagan Administration in 1981, is moving towards less government control over technological side effects. The technology-related regulatory agencies like the Environmental Protection Agency, the Occupational Health and Safety Administration, and the Nuclear Regulatory Committee, created in the 1970s in response to growing public concern, are going to see their activities subjected to centralized White House control through coordinating mechanisms like the Office of Management and the Budget and the Council of

188 QUESTIONING TECHNOLOGY

Economic Advisors. The net effect will be to shift technology control of this type from the public to the private sector. Furthermore, there will be increasing applications of the quantitative cost/benefit standard, and greater weight will be assigned to the goal of corporate profits in the private sector than to the concern for environmental integrity, urban health and safety, and consumer protection in the public sector.[7]

Self-Help: A Viable Alternative in the Absence of Controls?

If the values of Technological Nihilism continue to prevail, if there are no meaningful efforts to rethink technology and create soft and appropriate technologies, and if no effective controls on technological innovation are established, then short of participation in a revolution which seeks to transform our nihilistic technological society, self-help in individual cases may remain the only recourse for most citizens. For example, lumber companies cancelled their plans to spray timberland in Northern California with phenoxy herbicides 2,4-D and 2,4,5-T after irate citizens threatened to shoot down the helicopter which was doing the spraying.[8] These chemicals were used in Vietnam by US forces and were condemned by environmentalists as toxic to wildlife and as a cause of human birth defects.

In France, three discs of dangerous radioactive nuclear wastes, stolen from a nuclear waste facility, were hidden under the seat of a car owned by a middle-management executive at the plant. It was not clear at the time the press reported this story, however, whether the hiding of the deadly discs was intended as a callous prank or as a direct and radical means of discouraging the application of a dangerous and controversial technology, nuclear energy.[9]

Farmers in Minnesota tried to obstruct work crews who were assembling towers for a high-voltage transmission line which the farmers believed would expose them to dangers of high voltage and ozone release. Their efforts, however, were unsuccessful, because the governor called out state troopers.[10]

Local fishermen on the island of Kamishima, Japan, concerned about detergent pollution of waters containing fish and abalone, formed a "pollution vigilante" group. They visited each household, confiscating detergents and toothpaste and giving residents old-fashioned soap and soap-powder (and counselling the use of salt instead of toothpaste) in exchange. In this example of self-help there was no violence on the part of the government or the citizenry.

Residents of the island, a cohesive community, reportedly cooperated fully with the vigilantes efforts.[11]

Citizen self-help, however, whether peaceful or violent, generally will provoke counter-responses from business corporations and the state which protects their interests and possesses a coercive superstructure of courts, the police, the national guard, and prisons. As far as the private sector is concerned, the surveillance activities of utility companies worried about activists who have organized to oppose rate increases and the construction of nuclear power plants are indicative of what we may expect. In some areas, utilities have photographed participants in peaceful, constitutionally-protected nuclear and rate-increase protest rallies, have identified protestors by recording license-plate numbers (this data was sent, unsolicited, to the federal government's Nuclear Regulatory Commission), and have compiled dossiers and files on known activists.[12] Increasingly, private corporations may organize their own COINTELPRO-type operations,[13] establishing, in effect, their own secret police and spy units.

French anti-nuclear activists have a slogan, "A Nuclear State is a Police State," which suggests governments will resort to more repressive measures as dangerous and controversial technologies like nuclear power are more widely utilized. Governments (and private industry) are concerned in the case of nuclear energy with the problem of possible theft of plutonium and the sabotage or take-over of a plant by terrorists. Furthermore, there is also the problem of political unrest as activism by opponents of nuclear power increases. The French anti-nuclear slogan suggests that the state's repressive mechanisms will be brought to bear on these individuals as well. Today, reactors themselves are equipped with electronic sensors like those developed in Vietnam and are patrolled by security guards and police dogs. As the fears of the nuclear industry increase, it is possible that inhabitants of communities near nuclear facilities will be subjected to closer police surveillance, including wiretapping of telephones and other types of electronic snooping. Important details of their private, professional, and political lives may be collected in special computerized data banks related to "nuclear security".

It is clear that in addition to the dangers of social and ecological breakdown, the continuation of a society based on Technologial Nihilism means increasing political oppression. Technological Nihilism and technological fascism will go hand-in-hand; and the more desperate citizens resort to self-help against uncontrolled technologies which are harmful to the person and the planet, the more this tendency toward a technologically sophisticated police-state society will be reinforced. This is another reason why we must restore a civilization of limits, before it is too late.

1 Michel Bosquet, "Mensonges par omission", *Le Nouvel Observateur*, 21 April 1975, p.45.
2 Schwartz, pp. 225-237.
3 *Palo Alto Times*, 17 February 1975.
4 Since 1963, the US has had eighteen "accidental ventings" at the Nevada test site. *San Francisco Sunday Examiner and Chronicle*, 7 January 1979.
5 *San Francisco Sunday Examiner and Chronicle*, 7 January 1979.
6 Commoner, *The Closing Circle*, pp. 146-147.
7 Schwartz, p. 235.
8 *San Francisco Chronicle*, 2 April 1978.
9 *San Francisco Sunday Examiner and Chronicle*, 27 February 1977.
10 *San Francisco Chronicle*, 20 January 1977; *Wall Street Journal*, 10 October 1973.
11 *Wall Street Journal*, 4 November 1974.
12 *San Francisco Chronicle*, 6 November 1977.
13 *San Francisco Sunday Examiner and Chronicle*, 1 June 1980. "Interview with Lois Gibbs, Chair of Love Canal Homeowners' Association"; *CED News*, April 1980; *Wall Street Journal*, 12 September 1980.

Politics and the Restraint of Science

LEONARD A. COLE

Anthony Standen wrote in 1956 that the public regarded scientists as "lofty and impeccable . . . Since it is only natural to accept such flattery, the scientists accept the layman's opinion about themselves."[1] The scientist whose formative professional years occurred during World War II and the decade following is unlikely now to shed the image of grandeur about science and scientists imbedded earlier. On the other hand, those who are now in their twenties and thirties were exposed to far more skepticism about science. As a result, they may have developed different attitudes from their predecessors.

Observers less committed to the earlier ethos, including the younger scientists, may now more easily sense its implausibility. They are likely to notice the contradiction that scientific truth was supposedly unrelated to human values, but that discovering these truths was inherently good. "If you are a scientist you believe that it is good to find out how the world works; that it is good to find out what the realities are," said Robert Oppenheimer in 1945.[2] "Good" is a value judgement, and if scientific understanding is good, then understanding science has been inexorably tied to human values.

This developing recognition may also be linked to views about the conduct of scientific activity and to the appropriate role of political authority. The argument is consistent with Martin Brown's observation that "it is the young scientist who is most aware of the failure of science and most willing to do something about it."[3] The more one recognizes the human relationship to scientific activity, the greater the sense of responsibility for its consequences, harmful as well as beneficial. Increased receptivity to governmental intervention should therefore not be surprising because the government is the ultimate instrument devised by society to protect itself from danger.

Implications for the future

If the pattern of responses by scientists according to age is largely based on generational experiences, as suggested here, several consequences may be anticipated. As the public's earlier view of the grandeur of scientists helped shape the attitudes of scientists, so will continuing public skepticism also influence the self-image of future scientists. An increase in the proportion of scientists who reject the traditional assumptions and values may therefore be expected. In attitudes about the three categories — the nature of science, specific activities, governmental interference — the number of scientists whose positions parallel those of the youngest scientists in this study will continue to grow.

If more scientists relate science and its understanding to human values, scientific hazards will be seen increasingly as human responsibilities. In consequence, as with younger scientists in this sample, the scientific comunity may become less antipathetic to governmental regulation. The trend in each of the categories dealt with in the survey appears to be moving in that direction. In addition, as more scientists assimilate the new values, they will reinforce public consciousness of the hazards as well as the benefits of science and technology, thus increasing public sensitivity and skepticism as experts confess their own uncertainties.

Before an equilibrium is reached, continuing ferment and skepticism among the public and within the scientific community is to be expected. Any equilibrium will likely rest on the notion that scientific activity should be subject to contraints no greater and no less than for other activities that affect the public welfare.

The proper role of science — a competing interest

Neither abnegation nor overbearance on the part of government is desirable. George Ball's dictum that scientists should "decide what research to undertake and how to undertake it, subject only to such safeguards as they might individually or collectively impose,"[4] can hardly be justified in view of experiences cited in this book. Nor is there comfort in Peter Hutt's stifling admonition that "the role of scientists should be limited to conducting the type of research that the public concludes to be relevant to its needs (like the role of generals in conducting wars)."[5]

Rather we should understand that scientists comprise one among many interest groups. Many observers shrink from this recognition. Some have referred to science as a sacred establishment and to scientists as a priesthood or ecclesiastical order.[6] Neither science nor scientists (nor anyone else) deserve such an exalted position in a democracy. The scientific community is an interest group that should have to compete in the political and economic marketplace as do lawyers, teachers, and businessmen. Despite entreaties by some, scientists should no sooner be the exclusive regulators of science than farmers should be of agriculture or company executives of business.

Though this idea is unsettling to those who seek neat, circumscribed answers, dealing with science in a democratic society cannot be tidy. Pluralist politics is unavoidably imprecise. As Alexander Rich recognized at a symposium on managing science, if the problems are handled "in a pluralist manner, with many different groups having control of particular sectors in which thay are interested, we will in fact muddle through in a way that is very much consonant with our democratic principles."[7]

But "muddling through" does not preclude the need for occasional innovation. Whether through a Science Hearings Panel as proposed in this book, or in some similar manner, a systematic means to expose potentially disastrous scientific activity seems essential. As scientific and technological activities continue to multiply, this need becomes more compelling. The structures and values that underlie the American political system are suited to the purposes of such a panel. Opennesss, an educated polity, and a commitment to the people's welfare are central to the American political culture. To the extent that a panel would foster these attributes, the citizenry would be better protected from a variety of excesses related to science — from externally imposed truth, from unethical research, from masking the dangers of certain activities.

The threat of irreversible hazards may best be dealt with when understood as another in a long list of threats to liberty. Protection from them will require no less vigilance than has been required against abuse of individual liberties since the founding of the nation. No miracle institutions can insure protection from such awesome threats. Contrary to what Willard Libby said when he was a member of the Atomic Energy Commission, the American (and the world's) population should not have to learn to live with radioactive fallout. We do, however, have to live with its threat.

The thought is sobering but not hopeless. Neither science nor scientists will save us from externally imposed scientific truth or from irreversible hazards. Only the nexus of structures and values that comprises the political system can. Not regulatory agencies, democratic elections, enlightened citizens, a free press, a Science Hearings Panel, or any other single institution can insure protection. They are all essential.

Predicting the future of the relationship between science and politics is particularly risky amid the ferment that attends the issue today. Our analysis encourages confidence that the American polity is immune to perversions on the order of the historic cases. Unfortunately the frequency with which these incidents are linked by scientists and scholars to events in America blurs the issue that should be addressed: whether and how the political system can protect us against science and technology that threaten society with catastrophe.

The need for governmental interference, increasingly acknowledged within the scientific community, demonstrates a willingness by experts to question outworn assumptions. While many scientists cling to the traditional notion that science and scientists should remain unrestrained, others recognize changed conditions. The threat of irreversible hazards, the potential dangers in some activities, and public concern have all served to heighten the sensitivity of many scentists to issues that they had previously ignored. We may hope that this emerging awareness among scientists and the public will spur the national effort to confront the problems.

1 Anthony Standen, *Science is a Sacred Cow*, (New York: E. P. Dutton & Co., 1956), pp. 14-15.

2 Alice Kimball Smith and Charles Weiner, *Robert Oppenheimer, Letters and Recollections*, (Cambridge, Massachusetts: Harvard University Press, 1980), p. 317. More recently Phillip Handler, while president of the National Academy of Sciences, chided all who doubt the assumption that "objective knowledge is an unquestioned good." *Science*, vol. 208, no. 4448 (June 6, 1980), p. 1093.

3 Martin Brown in *The Social Responsibility of the Scientist*, Martin Brown ed. (New York: The Free Press, 1971), p. 271.

4 Quoted in Sissela Bok, "Freedom and Risk" *Daedalus*, vol. 107, no. 2 (spring 1978), p. 118.

5 Peter Barton Hutt, "Public Citizens of Health Service Policy" in *ibid.*, p. 163.

6 Standen, *Science is a Sacred Cow*; Don K. Price, *The Scientific Estate* (New York: Oxford University Press, 1965); Ralph E. Lapp, *The New Priesthood* (New York: Harper & Row. 1965). The authors, however, generally expressed wariness about popular acceptance of science in such lofty terms.

7 In "Commentaries from a Series of Academy Forums" *Science: An American Bicentennial View* (Washington DC: National Academy of Sciences, 1977), p. 75.

Saturn and Scientism

T. FULANO

There she is*, looking vaguely pornographic on the glossy covers of the weekly magazines, the planet Saturn. What have we discovered? I don't know, I haven't read them, feeling squashed as I do to the Earth by the giddying inertia of this century which plummets like a flaming satellite towards the nothingness.

Grey skies, the weather turning cold, sirens in the distance. Some citizens walk by whispering reverently of the wonders of Saturn, disputing the number of rings and moons according to the latest counts, as the corroding universe about them threatens to be annihilated. They drool over photographs of a planet most of them couldn't spot in a clear night sky — that is, if the night sky hadn't already been colonized and obliterated by the city light and the lethal dust of the very civilization which made it possible to send gadgets and technicians to the stars. But everything is so groovy on Saturn, so colorful and tempestuous. They know because they watched it all on television.

The spectacularization of space: a form of disassociation from an Earth which is coming apart at the seams. It is an odd addiction to scientific trivia which on the one hand resembles the "equipment mania" of hobbyism, and a futuristic brand of scientistic and mystical obscurantism on the other. They are hypnotized by the latest additions to an endless series of eternally changing scientistic truths. They think the universe is "out there" in the sky. They turn it into an alienated object just as they turn their daily creative activity and their life energy into commodities. And well they should so that the existing state of affairs may continue to prevail, for if they realized that *this* is their universe, that the only universe is their own world and their own lives, and that they are converting it into a vast, radioactive ruin, they would tremble with fear, and with passion and rage, and they would overturn it all.

They think going great distances at enormous velocities will bring them understanding. But the mystery is all around them; the center

*See "Questioning Technology, Questioning Patriarchy," p. 1.

of the universe is everywhere, and the circumference nowhere. They are on a treadmill, no matter what the velocity of their machines, and when they arrive at the remote reaches that they seek, they understand no more than they did when they left. (They could have gone there more readily in dreams.)

They think that if they launch enough rockets and feed enough facts into computers that they will someday discover the secret to life. But they will never find any answers, only more facts to feed to their computers, truths which will quickly be overthrown by new, more "revolutionary" truths. They have mistaken the question of *how to live* in their universe, with a technological scholasticism. They will only succeed in creating a world of gadgets, gadgets which will do their living for them. What does it mean, after all, to vicariously explore a pseudo-universe by gazing at the video-image on a television screen? What is this knowledge but a massive, technocratic illusion?

They think that the space program represents more freedom, an extension of their universe. In reality it is an element in their reduction. The night sky no longer exists for modern humanity. We live in these cities like rats in holes. Primitives, on the other hand, have a much more intimate relation with the stars, which haven't been reduced to mere images. They know the seasons and the plants, guide their migrations and their agricultural cycles by them; they name the heavenly bodies, feel their power. Modern humanity's relation with the stars is mediated by experts; we look at the planetary pornography in magazines, and know a sampling of reified facts and scientistic gibberish which we can spout when it is convenient, but we may as well be living in a cave. Our universe is an artifice, and nature is a commodity to be consumed. Our waking life is reduced to the cubicle and the production line, our dreams to television static, our wisdom to statistical jargon. Modern humanity thinks its scientist metaphors more accurately reflect "reality" than the mystic metaphors of the primitive, but they only reflect an impoverishment and a fundamental widespread ignorance.

The defenders of the scientist faith cite the spirit of discovery and exploration in their ecstatic panegyrics for the space program.The hallucination which these addicts of technology defend is nothing more than the spirit of capitalism in its early development, which is linked historically with the exploration and conquest of the "new world", a commercial expansion which initiated the greatest pillage and slaughter of all time. The "achievements" of capital cannot be considered separately from the slavery and extermination of indigenous peoples everywhere. The space program has been paid for in oceans of blood.

But of course it won't be "Man" who explores and conquers outer space any more than it was "Man" who explored and conquered America, but Capital. Human beings will function as the pawns of the military-industrial complex, East and West, which will transform space into capital. Incredible profits will be reaped in space, perhaps as it is divided into territories and spheres of influence. And if there are profits to be made, they will certainly lead to imperialist wars in the skies, star wars which will inevitably spread to the Earth.

This sinister side of the space program reveals how far the disassociation can go. *Everyone knows* that every advance in space technology and computer technology which accompanies it is an advance in military technology. Faster missiles, more accurate trajectories, more durable equipment, more efficient fuels, intensify the arms race and bring us closer to holocaust. *Everyone* knows this, but few utter it aloud.

The adventure is here and now on earth, not in desperate technological forays into space. We must begin by taking back this world which has been stolen from us, and learn to live upon it so that we can stand beneath the stars on a clear and silent night and know who we truly are and where we are going.

10. Is technology "neutral"?

Interest in digital watches seems to be waning. Concert-goers are no longer greeted by signs that ask them to turn of the "beep beep". In fact, the chorus of "beep beeps" that used to occur on the hour is now a sign that one is surrounded by the sort of folks that used to carry slide rules in their back pockets.

The digital watch, like a sextant, gives you your position *now*. But it does not measure or predict. A person with the clockface watch has the edge, somehow. And so, slowly but surely, the drift has been back to "telling time".

If time is alienating, high tech can increase the estrangement. But only with our assistance. The riddle is solved: guns don't kill people; people sometimes kill people; mostly, people kill people *with guns*.

The Zerzans find in the invention of the factory system a conscious tactic of social engineering, while ex-ad {man}* Mander characterizes television as fundamentally alienating and hence unreformable. Ian Reinecke demonstrates a deadening fragmentation of work that is being heightened, and not accidentally, by high tech work processes. Our shortest selection is Ellul's elucidation of an autonomous technology whose boundaries have grown beyond any reasonable definition of neutrality.

*See "Questioning Technology. Questioning Patriarchy", p. 1.

Industrialism and Domestication

JOHN AND PAULA ZERZAN

The independent craftsman* was a threatening adversary to the employing class, and he clung strongly to his prerogatives: his well-known propensity, for instance, to reject "the higher material standard of the factory towns," in Thompson's phrase, to gather his own fruits, vegetables and flowers, to largely escape the developing industrial blight and pollution, to gather freely with his neighboring workers at the dinner hour. Thompson noted a good example of the nature of the domestic worker in "the Yorkshire reputation for bluntness and independence" which could be traced to what local historian Frank Peel (early 19th century) saw as "men who doffed their caps to no one, and recognized no right in either squire or parson to question or meddle with them."

Riots, tumults and disorders

Turning to some of the specifics of pre-factory system revolt in England, the following from Ashton provides a good introduction:

Following the harvest failure of 1709 the keelmen of the Tyne took to rioting. When the price of food rose sharply in 1727 the tin miners of Cornwall plundered granaries at Falmouth, and the coal miners of Somerset broke down the turnpikes on the road to Bristol. Ten years later the Cornish tinners assembled again at Falmouth to prevent the exportation of corn, and in the following season there was rioting at Tiverton. The famine of 1739-40 led to a 'rebellion' in Northumberland and Durham in which women seem to have taken a leading part: ships were boarded, warehouses broken open, and the Guld at Newcastle was reduced to ruins. At the same time attacks on corn dealers were reported from North and South Wales. The years 1748 and 1753 saw similar happenings in several parts of the country; and in 1756-7 there was hardly a county from which no report reached the Home Office of the pulling down of corn mills or Quaker meeting houses or the rough handling of bakers and grain dealers. In spite of drastic penalties the same thing occurred in each of the later dearths of the century: in 1762, 1765-7, 1774, 1783, 1795 and 1800.

*See "Questioning Technology, Questioning Patriarchy," p.1.

This readiness for direct action informs the strife in textiles, the industry so important to England and to capitalist evolution, where, for example, "discontent was the prevalent attitude of the operatives engaged in the wool industries for centuries," said Burnley in his *Historys of Wool and Woolcombing*. Popular beliefs give ample evidence to this, as does the case of rioting London weavers, who panicked the government in 1675. Lipson's *History of the Woollen and Worsted Industries* provides many instances of the robustness of domestic textile workers' struggles, including that of a 1728 weavers' strike which was intended to have been pacified by a meeting of strike leaders and employers; a "mob" of weavers "burst into the room in which the negotiations were taking place, dragged back the clothiers as they endeavored to escape from the windows, and forced them to concede all their demands." Or these additional accounts by Lipson:

The Wiltshire weavers were equally noted for their turbulent character and the rude violence with which they proclaimed the wrongs under which they smarted. In 1738 they assembled together in a riotous manner from the villages round Bradford and Trowbridge, and made an attack upon the house of a clothier who had reduced the price of weaving. They smashed open the doors, consumed or spoiled the provisions in the cellar, drank all the wine they could, set the casks running, and ended up by destroying great quantities of raw materials and utensils. In addition to this exploit they extorted a promise from all the clothiers in Melksham that they would pay fifteen pence a yard for weaving ... Another great tumult occurred at Bradford (Wiltshire) in 1752. Thirty weavers had been committed to prison; the next day above a thousand 'weavers assembled, armed with bludgeons and firearms, beat the guard, broke open the prison, and rescued their companions.

Similarly, J. P. Kay was driven from Leeds in 1745 and from Bury in 1753, as outbreaks of violence flared in many districts in response to his invention, the flying shuttle for mechanizing weaving.

Wadsworth and Mann found the Manchester Constables Accounts to have reported "great riots, tumults and disorders" in the late 1740s, and that "after 1750 food riots and industrial disputes grew more frequent," with outbreaks in Lancashire (the area of their study) virtually every year. These historians further recount "unrest and violence in all parts of the country" in the middle to late 1750s, with Manchester and Liverpool frequently in alarm and "panic among the propertied classes."

After sporadic risings, such as Manchester in 1762, the years 1764-68 saw rioting in almost every county in the country; as the

King put it in 1766, "a spirit of the most daring insurrection has in diverse parts broke forth in violence of the most criminal nature." Although the smashing of stocking frames had been made a capital offense in 1721, in a vain attempt to stem worker violence, Hobsbawm counted 24 incidents of wages and prices being forcibly set by exactly this type of riotous destruction in 1766 alone.

Sporadic rioting occurred in 1769, such as the anti-spinning jenny outbursts which menaced the inventor Hargreaves and during which buildings were demolished at Oswaldthistle and Blackburn in order to smash the hated mechanization. A whole new wave began in 1772. Sailors in Liverpool, for example, responded to a wage decrease proposal in 1775 by "sacking the owners' houses, hoisting 'the bloody flag', and bringing cannon ashore which they fired on the Exchange," according to Wadsworth and Mann.

The very widespread anti-machinery risings of 1779 saw the destruction of hundreds of weaving and spinning devices which were too large for domestic use. The rioters's sentiments were very widely shared, as evidenced by arrest records that included miners, nailmakers, laborers, joiners — a fair sample of the entire industrial population. The workers' complaint averred that the smaller machines are "in the hands of the poor and the larger 'patent machines' in the hands of the rich," and "that the work is better manufactured by small [textile machines] than by large ones."

This list, very incomplete as it is, could be easily extended into the many early 19th century outbreaks, all of which seem to have enjoyed great popular support. But perhaps a fitting entry in which to close this sample would be these lines from a public letter written by Gloucestershire shearmen in 1802: "We hear in Formed that you got Shear in mee sheens and if you Don't Pull them Down in a Forght Nights Time we will pull them Down for you Wee will you Damd infernold Dog."

The factory and social control

This brief look at the willfulness of the 18th century proletariat serves to introduce the conscious motivation behind the factory system. Sidney Pollard (*The Genesis of Modern Management*) recognized the capitalists' need of "breaking the social bonds which had held the peasants, the craftsmen and the town poor of the eighteenth century together in opposition to the new order." Pollard saw too the essential nature of the domestic system, that the masters "had to depend on the work performed in innumerable tiny domestic workshop units, unsupervised and unsupervisable. Such incompatibility," he concluded, "was bound to set up tensions and to drive the merchants to seek new ways of production, imposing their own managerial achievements and practices in the productive sector."

This underlying sense of the real inadequacy of existing powers of control was also firmly grasped by David Landes (*The Unbound Prometheus*): "One can understand why the thoughts of employers turned to workshops where the{men}would be brought together to labour under the watchful overseers, and to machines that would solve the shortage of{manpower}while curbing the insolence and dishonesty of the men." According to Wadsworth and Mann, in fact, many employers definitely felt that "the country would perish if the poor — that is, the working class — were not brought under severe discipline to habits of industry and docile subordination."

Writing on the evolution of the 'central workshop' or factory, historian N. S. B. Gras saw its installation strictly in terms of control of labor: "It was purely for purposes of discipline, so that workers could be effectively controlled under the supervision of{foremen."} Factory work itself became the central weapon to force an enemy character into a safe, reliable mold following the full realization that they were dealing with a recalcitrant, hostile working class whose entire morale, habits of work and culture had to be broken. Bowden described this with great clarity: "More directly as a result of the introduction of machinery and of large-scale organization was the subjection of the workers to a deadening mechanical and administrative routine."

Richard Arkwright (1732-1792) agreed completely with those who saw the need for consciously spurring consumption, "as to the necessity of arousing and satisfying new wants," in his phrase. But it is as the developer of cotton spinning machinery that he deserves a special word here; because he is generally regarded as the most prominent figure in the history of the textile industries and even as 'the founder of the factory system.' Arkwright is a clear illustration of the political and social character of the technology he did so much to advance. His concern with social control is very evident from his writings and correspondence, and Mantoux (*The Industrial Revolution in the Eighteenth Century*) discerned that "His most original achievement was the discipline he established in the mills."

Arkwright also saw the vital connection between work discipline and social stability: "Being obliged to be more regular in their attendance on their work, they became more orderly in their conduct." For his pioneering efforts, he received his share of appropriate response; Lipson relates that in 1767, with "the news of the riots in the neighbourhood of Blackburn which had been provoked by Hargreaves' spinning jenny," he and his financial backer Smolley, "fearing to draw upon themselves the attention of the machine-wreckers, removed to Nottingham." Similarly, Arkwright's Birkacre mill was destroyed by workers in 1779. Lipson ably summarizes his managerial contribution:

In coordinating all the various parts of his vast industrial structures; in organising and disciplining large bodies of men, so that each man fitted into his niche and the whole acted with the mechanical precision of a trained army . . . in combining division of labour with effective supervision from a common centre . . . a new epoch was inaugurated.

Andrew Ure's *Philosophy of Manufactures* is one of the major attempts at an exposition of the factory system, a work cited often by Marx in *Capital*. Its revealing preface speaks of tracing "the progression of the British system of industry, according to which every process peculiarly nice, and therefore liable to injury from the ignorance and waywardness of workmen is withdrawn from handicraft control, and placed under the guidance of self-acting machinery." Examining the nature of the new system, then, we find, instead of domestic craft labor, "industrial labor . . . [which] imposes a regularity, routine, and monotony . . . which conflicts . . . with all the inclinations of a humanity as yet unconditioned into it," in the words of Hobsbawm. Factory production slowly supplanted that of the domestic system in the face of fierce opposition (discussed below), and workers experienced the feeling of daily entering a prison to meet the new "strain and violence" of work, as the Hammonds put it. Factories often resembled pauper workhouses or prisons, after which they had actually often been modelled; Max Weber saw a strong initial similarity between the modern factory and the Russian serf-labor workshops, wherein the means of production and the workers themselves were appropriated by the masters.

Hammonds' *Town Labourer* saw "the depreciation of human life" as the leading fact about the new system for the working classes: "The human material was used up rapidly; workmen were called old at forty." Possibly just as important was the novel, "inhuman" nature of its domination, as if all "were in the grasp of a great machine that threatened to destroy all sense of the dignity of human life," as the Hammonds described it. A famous characterization by

J. P. Kay (1832) put the everyday subjugation in hard to forget terms:

Whilst the engine runs the people must work — men, women and children are yoked together with iron and steam. The animal machine — breakable in the best case, subject to a thousand sources of suffering — is chained fast to the iron machine, which knows no suffering and no weariness.

Resistance to industrial labor displayed a great strength and persistance, reflecting the latent anti-capitalism of the domestic worker who was "the despair of the masters" in a time when a palpable aura of unfreedom clung to wage-labor. Lipson tells us, for example, of Ambrose Crowley, perhaps the very first factory owner and organizer (from 1691); that he showed an obsession with the problem of disiplining his workers to "an institution so alien in its assumptions about the way in which people should spend their lives."

Lewis Paul wrote from his London firm in 1742 that "I have not half my people come to work today and I have no fascination in the prospect that I have to put myself in the power of such people." In 1757 Josiah Tucker noted that factory-type machinery is highly provocative to the populace who "never fail to break out into riots and insurrections whenever such things are proposed." As we have seen, and as Christopher Hill puts it, "Machine-breaking was the logical reaction of free ⦃men⦄ . . . who saw the concentration of machinery in factories as the instrument of their enslavement."

A hosiery capitalist, in admitting defeat to the Committee on Woollen Manufacture, tells us much of the independent spirit that had to be broken:

I found the utmost distaste on the part of the ⦃men⦄ to any regular hours or regular habits . . . The ⦃men⦄ themselves were considerably dissatisfied, because they could not go in and out as they pleased, and go on just as they had been used to do . . . to such an extent as completely to disgust them with the whole system, and I was obliged to break it up.

The famous early entrepreneurs, Boulton and Watt, were likewise dismayed to find that the miners thay had to deal with were "strong, healthy and resolute ⦃men,⦄ setting the law at defiance; no officer dared to execute a warrant against them."

Wedgewood, the well-known pottery and china entrepreneur, had to fight "the open hostility of his work-people" when he tried to develop division of labor in his workshops, according to Mantoux.

And Jewitt's *The Wedgewoods*, exposing the social intent of industrial technology, tells us "It was machinery [which] ultimately forced the worker to accept the discipline of the factory."

The threat of job security

Considering the depth of workers' antipathy to the new regimen, it comes as no surprise that Pollard should speak of "the large evidence which all points to the fact that continuous employment was precisely one of the most hated aspects of factory work." This was the case because the work itself, as an agent of pacification, was perceived 'precisely' in its true nature. Pollard later provides the other side of the coin to the workers' hatred of the job; namely, the rulers' insistence on it for its own (disciplinary) sake: "Nothing strikes so modern a note in the social provisions of the factory villages as the attempts to provide continuous employment."

Returning to the specifics of resistance, Sir Frederic Eden, in his *State of the Poor* (1797), stated that the industrial laborers of Manchester "rarely work on Mondays and that many of them keep holiday two or three days in the week." Thus Ure's tirades about the employees' "unworkful impulses," their "aversion to the control and continuity of factory labor," are reflected in such data as the fact that as late as 1800, spinners would be missing from the factories on Mondays and Tuesdays. Absenteeism, as well as turnover, then, was part of the syndrome of striving to maintain a maximum of personal liberty.

Max Weber spoke of the "immensely stubborn resistance" to the new work discipline, and a later social scientist, Reinhard Bendix, saw also that the drive to establish the management of labor on "an impersonal systematic basis" was opposed "at every point." Ure, in a comment worth quoting at length, discusses the fight to master the workers in terms of Arkwright's career:

The main difficulty [he faced was] above all, in training human beings to renounce their desultory habits of work, and to identify themselves with the unvarying regularity of the complex automaton. To devise and administer a successful code of factory discipline, suited to the necessities of factory diligence, was the Herculean enterprise, the noble achievement of Arkwright. Even at the present day, when the system is perfectly organized, and its labour lightened to the utmost, it is found nearly impossible to convert persons past the age of puberty, whether drawn from rural or from handicraft occupations, into useful factory hands.

We also encounter in this selection from Ure the reason why early factory labor was so heavily comprised of the labor of children, women, and paupers threatened with loss of the dole. Thompson quotes a witness before a Parliamentary investigative committee, that "all persons working on the power-loom are working there by force because they cannot exist any other way." Hundreds of thousands clung to the deeply declining fortunes of hand-loom weaving for decades, in a classic case of the primacy of human dignity, which Mathias (*The First Industrial Nation*) notes "defied the operation of simple economic incentives."

What Hill termed the English craftsmen's tradition "of self-help and self-respect" was a major source of that popular will which denied complete dominion by capital, the "proud awareness that voluntarily going into a factory was to surrender their birth-right."

Thompson demonstrates that the work rules "appeared as unnatural and hateful restraints" and that everything about factory life was an insult. "To 'stand at their command' — this was the most deeply resented indignity. For he felt himself, at heart, to be the real maker of the cloth . . . "

This spirit was why, for example, paper manufacturers preferred to train inexperienced labor for the new (post-1806) machine processes, rather than employ skilled hand paper makers. And why Samuel Crompton, inventor of the spinning mule, lamented, relatively late in this period:

To this day, though, it is more than thirty years since my first machine was shown to the public, I am hunted and watched with as much never-ceasing care as if I was the most notorious villain that ever disgraced the human form; and I do affirm that if I were to go to a smithy to get a common nail made, if opportunity offered to the bystanders, they would examine it most minutely to see if it was anything but a nail.

The battle raged for decades, with victories still being won at least as late as that over a Bradford entrepreneur in 1882, who tried to secretly install a power-loom but was discovered by the domestic workers. "It was therefore immediatley taken down, and placed in a cart under a convoy of constables, but the enraged weavers attacked and routed the constables, destroyed the loom, and dragged its roller and warp in triumph through Baildon." Little wonder that Ure wrote of the requirement of "a Napoleon nerve and ambition to subdue the refractory tempers of work-people."

Mental mutilation

Without idealizing the earlier period, or forgetting that it was certainly defined by capitalist relationships, it is also true, as Hill wrote, "What was lost by factories and enclosures was the independence, variety and freedom which small producers had enjoyed." Adam Smith admitted the "mental mutilation" due to the new division of labor, the destruction of both an earlier alertness of mind and a previous "vivacity of both pain and pleasure."

Robert Owen likewise discussed this transformation when he declared, in 1815, that "The general diffusion of manufactures throughout a country generates a new character . . . an essential change in the general character of the mass of the people." Less abstractly, the Hammonds harkened back to the early 19th century and heard the "lament that the games and happiness of life are disappearing," and that soon "the art of living had been degraded to its rudest forms."

Four Arguments for the Elimination of Television

JERRY MANDER

Most Americans, whether on the political left, center or right, will argue that technology is neutral, that any technology is merely a benign instrument, a tool, and depending upon the hands into which it falls, it may be used one way or another. There is nothing that prevents a technology from being used well or badly; nothing intrinsic in the technology itself or the circumstances of its emergence which can predetermine its use, its control or its effects upon individual human lives or the social and political forms around us.

The argument goes that television is merely a window or a conduit through which any perception, any argument or reality may pass. It therefore has the potential to be enlightening to people who watch it and is potentially useful to democratic processes.

It will be the central point of this book that these assumptions about television, as about other technologies, are totally wrong.

If you once accept the principle of an army — a collection of military technologies and people to run them — all gathered together for the purpose of fighting, overpowering, killing and winning, then it is obvious that the supervisors of armies will be the sort of people who desire to fight, overpower, kill and win, and who are also good at these assignments: generals. The fact of generals, then, is predictable by the creation of armies. The kinds of generals are also predetermined. Humanistic, loving, pacifistic generals, though they may exist from time to time, are extremely rare in armies. It is useless to advocate that we have more of them.

If you accept the existence of automobiles, you also accept the existence of roads laid upon the landscape, oil to run the cars, and huge institutions to find the oil, pump it and distribute it. In addition you accept a sped-up style of life and the movement of humans through the terrain at speeds that make it impossible to pay attention to whatever is growing there. Humans who use cars sit in fixed positions for long hours following a narrow strip of gray pavement, with eyes fixed forward, engaged in the task of driving. As long as they are driving, they are living within what we might call "roadform". Slowly they evolve into car-people. McLuhan told us that cars "extended the human feet, but he put it the wrong way. Cars *replaced* human feet.

If you accept nuclear power plants, you also accept a techno-scientific-industrial-military elite. Without these people in charge, you could not have nuclear power. You and I getting together with a few friends could not make use of nuclear power. We could not build such a plant, nor could we make personal use of its output, nor handle or store the radioactive waste products which remain dangerous to life for thousands of years. The wastes, in turn, determine that *future* societies will have to maintain a technological capacity to deal with the problem, and the military capability to protect the wastes. So the existence of the technology determines many aspects of the society.

If you accept mass production, you accept that a small number of people will supervise the daily existence of a much larger number of people. You accept that human beings will spend long hours, every day, engaged in repetitive work, while supressing any desires for experience or activity beyond this work. The workers' behaviour becomes subject to the machine. With mass production, you also accept that huge numbers of identical items will need to be efficiently distributed to huge numbers of people and that institutions such as advertising will arise to do this. One technological process cannot exist without the other, creating symbolic relationships among technologies themselves.

If you accept the existence of advertising, you accept a system designed to persuade and to dominate minds by interfering in people's thinking patterns. You also accept that the system will be used by the sorts of people who like to influence people and are good at it. No person who did not wish to dominate others would choose to use advertising, or choosing it, succeed in it. So the basic nature of advertising and all technologies created to serve it will be consistent with this purpose, will encourage this behaviour in society, and will tend to push social evolution in this direction.

In all of these instances, the basic form of the institution and the technology determines its interaction with the world, the way it will be used, the kind of people who use it, and to what ends.

And so it is with television.

Far from being "neutral," television itself predetermines who shall use it, how they will use it, what effects it will have on individual lives, and, if it continues to be widely used, what sorts of political forms will inevitably emerge.

It was only after a long while and many half-steps of change in viewpoint that I finally faced the fact that television is not reformable, that it must be gotten rid of totally if our society is to return to something like sane and democratic functioning. So, to argue that case, especially considering that it involves a technology accepted as readily and utterly as electric light itself, is not something that ought to be done rapidly or lightly. Nor can such a case be confined to the technology itself, as if it existed aside from a context.

The first argument is theoretical and environmental. It attempts to set the framework by which we can understand television's place in modern society. Yet, this argument is *not* about television itself. In fact, television will be mentioned only occasionally. It is about a process, already long underway, which has successfully redirected and confined human experience and therefore knowledge and perceived reality. We have all been moved into such a narrow and deprived channel of experience that a dangerous instrument like television can come along and seem useful, interesting, sane, and worthwhile at the same time it further boxes people into a physical and mental condition appropriate for the emergence of autocratic control.

The second argument concerns the emergence of the controllers. That television would be used and expanded by the present powers-that-be was inevitable, and should have been predictable at the outset. The technology permits of no other controllers.

The third argument concerns the effects of television upon individual human bodies and minds, effects which fit the purposes of the people who control the medium.

The fourth argument demonstrates that television has no democratic potential. The technology itself places absolute limits on what may pass through it. The medium, in effect, chooses its own content from a very narrow field of possibilities. The effect is to drastically confine all human understanding within a rigid channel.

What binds the four arguments together is that they deal with aspects of television that are not reformable.

What is revealed in the end is that there is ideology in the technology itself. To speak of television as "neutral" and therefore subject to change is as absurd as speaking of the reform of a technology such as guns.

Electronic Illusions: A Skeptic's View of Our High Tech Future

IAN REINECKE

Neutralizing the threat

Neutrality is a claim made for the microchip almost as an instinctive defensive reaction by the technologists who have developed it or seek to introduce it. The following is a quote from an interview with one of the founders of a high-technology communications company in the United States

I think . . . about the question of the technology and how it is used — that the technology is absolutely neutral and the same microprocessors would be used for good or evil. The determination of that really comes through what the individual, what the collective group of individuals, align themselves with. What our values are, what priorities we have in life, I think that's what the real question is, not the technology.[1]

The added interest in those remarks is that they came from a man✻ who took the fortune he made in the computer industry and got out. Yet his questioning of social values and priorities did not lead him beyond the proposition that technology is neutral. He speaks as if it is a matter of individual preference about how we choose to implement the microchip, that its use will be determined by "society" and not by powerful groups and individuals within it. Any man or woman who works in an office or factory that has been invaded by the chip can tell him the truth. Working people are not generally consulted in the design or the introduction of computer technology. Their opinions are certainly not sought on the basic question or whether its introduction is necessary to do the work better. If they are lucky they have a trade union that ameliorates job loss or strives to maintain working conditions. As for the neutrality of the technology, they would be more convinced if there were a wealth of examples that contradicted their personal experience. Examples of technology that creates new jobs, maintains living

211

✻ See "Questioning Technology. Questioning Patriarchy," p. 1.

standards while reducing the working week, and increases their control over and autonomy at work are precious few.

The neutrality argument is no diversion in the new-technology debate. It expresses a central ideology succinctly in this quote: "Ethically, technology is neutral. There is nothing inherently either good or bad about it. It is simply a tool, a servant, directed and deployed by people for whatever purpose they want fulfilled." The people who wrote those words work for the corporate advertising agency of the international company United Technologies. Its subsidiaries include Pratt & Whitney, Mostek, Sikorsky Helicopters, and the Carrier Corporation, all high-technology companies. Alexander Haig, before he became Secretary of State in the Reagan administration, was its chairman. The quote is from a full-page advertisement run in many US and international business magazines under the heading "Technology's Promise."

United Technologies was preaching to what it regards as its natural audience in those advertisements, using publications that draw their readership from the middle and upper ranks of similar corporations. The purpose of rehearsing the argument in print for consumption by the faithful is to reinforce the message. Technology is good for business, and business is good for nations and for the people who live and work in them, the line runs. In many countries where technological development and diffusion are dominated by international companies operating little more than commercial shop windows, the argument that technology is neutral has little force. Those international corporate beneficiaries of local demand for technology export their profits, minimize their local tax obligations, and conceal their sales turnovers. They reveal meaningless profit figures, accept generous government subsidies, and return as little as possible to the national economy. The capacity to produce the technology does not in many instances flow to host countries and their citizens. Microelectronic technology is clearly good for them, but it is another question whether it is good for us.

The assertion that technology is neutral needs to be examined against examples of how it has been introduced. Consider this account of automation in a telephone exchange:

FADS (Force Administration Data System) is a computer system which, by measuring the pattern of traffic into a particular office each day, can anticipate employment needs for the next. Each morning, workers receive computer printouts listing their break and lunch times based on the anticipated traffic patterns for the day.

The FADS procedure has been in existence for a long time but recently it was computerized. Like other administrative systems, computer control squeezes out what little flexibility workers had before. Before computerization, a worker's morning break normally came about two hours after the beginning of the shift; now, it can come as early as fifteen minutes into the working day. Workers cannot go to the bathroom unless they find someone to take their place. "If you close your terminal," says Jean Miller, a service representative in Washington, "right away the computer starts clacking away and starts ringing a bell."[2]

Is the FADS system capable of being used equally for good or evil? Similar systems abound in offices: in word processors that measure productivity and telephone systems that trace calls to the extension from which they were made. Is its development an instance of the sort of progress the microchip has brought to working people? There is in FADS a disturbing similarity to the following account of working in a luggage manufacturing plant, with not a computer in sight:

You cannot at any time leave the tank. The pieces in the dye will burn while you're gone. If you're real, real sick and in urgent need, you do shut it off. You turn on the trouble light and wait for the toolman to come and take your place. But they'll take you to a nurse and check it out.[3]

Answering telephones is cleaner work than factory labor, but the conditions are transcended by that common principle of subservience to the working system, to the machine. That same principle is evident in this description of automation at Chrysler in Detroit:

A line I worked on moved so fast that they had a buzzer sound every time the line moved. When the buzzer sounded, you'd better move and move fast, or else you could get hurt bad. Behind me, just a few feet away, there was a water fountain. I wanted a swallow of water so bad and I thought maybe, if I worked as fast as I could in between the buzzer sounding, that I'd be able to jump back and get a drink of water. But no matter how much I tried, I could never get it. That swallow of water was so close, but it was like being on a desert.[4]

The common theme in the telephone exchange, the luggage factory, and the automobile plant is control — control of the process of doing or making things and of the human beings involved in it. People working in those production systems do not control them, but are "simply a tool, a servant, directed and deployed by people ... for whatever purpose they want," just as the United Technologies advertisement described the systems' non-human components. The people referred to are certainly not the people who act out their onerous roles as captives within the system. And the people cannot be the population at large, because they were not consulted through the ballot box or by any other means about how the technology should be deployed. The "people" who direct and control workers in production systems are members of the very powerful group in society that introduces the technology: the owners and managers of factories and offices.

A point omitted from much of the discussion about technology is that its purpose is built into the design. The systems that use electronics in offices, factories and homes are not designed as universally applicable tools, as good for one purpose as any other. They are not intended to serve a variety of ends. Primitive man may have used the same smooth rock for grinding crops, polishing bone, smoothing wood, shattering flint for spearheads, and smashing shells. Electronic systems narrow the focus of the machines to single, specific tasks.

Microelectronics is being used in machines that break down job skills into a series of simpler, more boring tasks. A book-keeping clerk's job a decade ago involved counting figures, calculating percentages, writing up ledgers, writing and typing letters, and sending them out. Counting can now be done on electronic calculators, or by programs on computers. A different program calculates payments and credits; word processors are used for typing; line printers turn out letters; address labels are printed out from electronically stored files. The office systems that take over manual work also formalize the control and monitoring that used to be done informally through direct supervision. They channel work through systems that also provide the means for assessing it. In some offices, bonus systems are tied to the number of times the keys on a keyboard are depressed each hour. The basis for reward is also the basis for punishment.

The values inherent in many electronics systems extend beyond offices and factories to reflect the society in which they exist. The systems emphasize the primacy of control and production geared to profit. They serve the interest of those who produce technology, those they sell it to, and those who benefit from it. The test of technology's supposed neutrality is whether its design is unaffected by the society around it. If it were really neutral, the technology of

FADS would still occur in a more democratic, equal society, where production demands and those of human beings were more evenly weighted.

The neutrality of a technology resides not in its theoretical possibility for good or evil but in how it is designed to be used. If in practice it is used only as a threat, as an instrument of control, as means of subjugation of many by few, its claim to neutrality is spurious.

Computer technology, in the way it has been introduced into many working environments in the last decade, is a newer version of Henry Ford's automative assembly line. The worker becomes simply one operating component within a larger process. Consider as an illustration a computerized typesetting company that produces magazines. A human being sits at a keyboard, in order to press the keys in a particular order. The keys transmit electronic signals to the machine. Screens display the words that have been typed. The computer it is connected to converts all the words into signals and processes them. A typesetting machine prints the words on strips of paper. Another human being pastes them up into pages, an intervention in the process that will be quite short-lived.

The value of the human being is as part of a whole, a component of the system, The designers of the technology have treated the worker as an equal with other components. It is an egalitarian approach; it treats the mechanical and human parts in the same way. Only if one is prepared to concede that workers are neutral can the technology be regarded as neutral. If that is not conceded, if one believes that human beings are not the passive, efficient, operating systems that their electronic comrades in harness are, the system they work in has no claims to neutrality.

1 Interview with Gene Richardson, a founder of Rolm, on "Four Corners" (Australian Broadcasting Commission, 1981).
2 Robert Howard, "Brave New Workplace" in *Working Papers*, November-December 1980, p. 26.
3 Grace Clements, luggage factory worker, quoted in Studs Terkel, *Working* (New York: Pantheon, 1974; London: Wildwood House, 1975).
4 Charles Denby, *Indignant Heart: Testimony of a Black American Worker* (Boston: South End Press, 1978; London: Pluto Press, 1979).

The Technological Society

JACQUES ELLUL

It was long claimed that technique was neutral. Today this is no longer a useful distinction. The power and autonomy of technique are so well secured that it, in its turn, has become the judge of what is moral, the creator of a new morality. Thus, it plays the role of creator of a new civilization as well. This morality — internal to technique — is assured of not having to suffer from technique. In any case, in respect to traditional morality, technique affirms itself as an independent power. Man alone is subject, it would seem, to moral judgement. We no longer live in that primitive epoch in which things were good or bad in themselves. Technique in itself is neither, and can therefore do what it will. It is truly autonomous.

However, technique cannot assert its autonomy in respect to physical or biological laws. Instead, it puts them to work; it seeks to dominate them.

*See "Questioning Technology, Questioning Patriarchy," p.1.

BIBLIOGRAPHICAL NOTE

The literature on "technology" (which is to say, on division of labor, production, industrialism and their impact on us and on nature) presents its own questions, such as why isn't it more developed in light of our present extremity. Must the planet, our lives on earth, become even more indigent, flattened out and distorted, leading up to complete destruction, before we can see technology for what it is? How much further can our dependency on technology increase, along with our estrangement from each other and the world, until we see it as a monstrous, non-neutral component of the death trip civilization is on?

It isn't that all this has been a total secret; in fact under very adverse circumstances the critique of the machine has begun to emerge. Friedrich Georg Jünger wrote his powerful *Failure of Technology* in the late 1930s in the belly of the National Socialist beast. Adorno and Horkheimer's *Dialectic of Enlightenment* is a creation of the immediate post-war '40s, under the shadow of Auschwitz. In the 1950s, a period of rampant industrial development, Jacques Ellul gave us *The Technological Society*, still the most thorough-going analysis of the techno-disease and its structural logic.

And yet in the late '80s, we are still slow to comprehend the depth of our alienation and oppression at the 'hands' of all-pervasive, artificial, galloping, polluting, de-eroticizing Industrial Civilization. Our enslavement to the machine has never been more complete, and this should be making it easier for us to see that freedom and wholeness are impossible until we undo division of labor, that our high tech societies rest more than ever on slavery in the mines and factories, computer-monitored office drudgery and poisonous ruin for nature, that real communication and real movement would mean a world without television, satellite hook-ups, jet-liners and cars.

217

Meanwhile, environmental activists seem content to tilt at superficial symptoms of the contagion, and marxists, despite Baudrillard's excellent *Mirror of Production*, continue to worship technological progress. But there are some signs that critical dialogue is alive and well. One standout is the American journal/newspaper *Fifth Estate*, which has done much since the late '70s to promote an examination of the subject at a basic level, mindful that to confront only the excesses of capital's manifestation as technology is to fail at the essential. And the initiative Freedom Press has shown with the book you are reading shows that these classical exponents of anarchy are aware of the importance of such a text, however modest its contribution.

We consider the passages in this anthology to constitute the best possible selections in answer to the ten questions we posed. Nonetheless, the reader should not assume that the works from which they are excerpted necessarily embody fundamentally negative assessments of technology.

INDEX